D0077429

Real-Time UML Workshop
for Embedded Systems

Real-Time UML Workshop
for Embedded Systems

by Bruce Powel Douglass, Ph.D.

AMSTERDAM • BOSTON • HEIDELBERG • LONDON
NEW YORK • OXFORD • PARIS • SAN DIEGO
SAN FRANCISCO • SINGAPORE • SYDNEY • TOKYO

Newnes is an imprint of Elsevier

ELSEVIER

Newnes

Newnes is an imprint of Elsevier
30 Corporate Drive, Suite 400, Burlington, MA 01803, USA
Linacre House, Jordan Hill, Oxford OX2 8DP, UK

Copyright © 2007, Elsevier Inc. All rights reserved.

No part of this publication may be reproduced, stored in a retrieval system, or transmitted
in any form or by any means, electronic, mechanical, photocopying, recording, or otherwise,
without the prior written permission of the publisher.

Permissions may be sought directly from Elsevier's Science & Technology Rights
Department in Oxford, UK: phone: (+44) 1865 843830, fax: (+44) 1865 853333,
e-mail: permissions@elsevier.com.uk. You may also complete your request on-line via the
Elsevier homepage (http://elsevier.com), by selecting "Support & Contact," then
"Copyright and Permission" and then "Obtaining Permissions."

 Recognizing the importance of preserving what has been written,
Elsevier prints its books on acid-free paper whenever possible.

Library of Congress Cataloging-in-Publication Data

Application submitted

British Library Cataloguing-in-Publication Data
A catalogue record for this book is available from the British Library.

ISBN-13: 978-0-7506-7906-0
ISBN-10: 0-7506-7906-9

For information on all Newnes publications
visit our website at www.books.elsevier.com

07 08 09 10 10 9 8 7 6 5 4 3 2 1

Printed in the United States of America

Working together to grow
libraries in developing countries

www.elsevier.com | www.bookaid.org | www.sabre.org

ELSEVIER   BOOK AID
            International   Sabre Foundation

This book is dedicated to my family—
two boys who make me proud,
a wife any man would envy,
and a stepdaughter who is so bright, she's scary.
It's good to be me—thanks, guys!

Contents

Preface

Most books on UML or real-time systems, even those that I have written, are what I call "lecture books." They might have exercises at the end of the chapters, but they are basically organized around "Here, let me tell you about that…" This book is fundamentally different. It is meant to be a workbook, a set of guided exercises that teach by practice and example, not by exposition. As such, we won't discuss in detail various aspects of the UML, or MDA, or even process. We will mention such things in passing, but the expectation is that the reader already has a reasonable understanding of the UML and wants practice and experience in the application of the UML to embedded and real-time systems development. This book is meant as a companion to another of my books, *Real-Time UML, Third Edition: Advances in the UML for Real-Time Systems.*[1] That book *is* a lecture book and explains, in some detail, the structural, behavioral, and functional aspects of the UML, particularly how it applies to the development of real-time and embedded systems.

The present book is organized primarily around two example problems, whose detailed problem statements can be found in the appendices. To give some breadth of experience in the exercises, both a small-scale and large-scale application are provided. The small-scale example is an intersection traffic-light control system. This is a small-scale system but is still complex enough to be a rich source of work for the reader.[2] The larger-scale example is an Unmanned Air Vehicle (UAV), although it is more properly called an Unmanned Combat Air Vehicle (UCAV). This system is highly complex and offers good experience in doing requirements analysis and architectural design for large-scale systems, something the traffic-light control system

---

[1] Douglass, Bruce Powel, *Real-Time UML, Third Edition: Advances in the UML for Real-Time Systems,* Addison-Wesley, 2004.

[2] One of the Laws of Douglass is "Anything is simple if you don't have to do it!" This certainly applies in spades to the development of real-time systems, since even "simple" systems can be very complex.

cannot. The point of these examples is to provide a context for the reader to explore the application of UML to the specification, analysis, and design of such systems. It is most assuredly *not* to provide complete designs of such systems. Such systems are created by teams (or, as in the case of the UCAV, teams of teams). If the purpose of the book was to provide complete designs of such systems, the book would be between 20 and 100 times its present size.

This book can be thought of as being composed of several parts. First is a brief overview of the basics—UML and process—provided in Chapters 1 and 2. The next section is a series of chapters that walk through the phases of an incremental spiral process—requirements analysis, object analysis, architectural design, mechanistic design, and detailed design—posing problems that occur in the context of the examples. The next section provides solutions for the problems posed in the previous section. Lastly, there is some end-matter—the appendices that provide the problem statements and a brief notational guide to the UML.

All the UML diagrams in this book are created in the Rhapsody™ tool from Telelogic, and a demo copy of Rhapsody is provided in this book. The full version offers validation, model execution, and production-quality code generation facilities, as well as interfaces to other tools, such as requirements management, XMI import/export, etc. While these features are enormously helpful in actual development, they are not relevant to the purpose of the book. If you prefer to use another tool for the exercises, or even (gasp) pen and paper, you can do that as well.

The solutions are, of course, the "pointy end" of the stick. They are meant, not so much as the absolute truth or measure of goodness in UML modeling, but instead as examples of solutions to the problem. If your solution differs from the one provided and still meets the solution criteria, then that's perfectly fine. There are always many good solutions to any problem.[3] If your solution differs from the provided solution, I recommend that you study the differences between the two and determine under which conditions one or the other would be preferable.

Audience

The book is oriented towards the practicing professional software developer and the computer science major, in the junior or senior year. It focuses on practical experience by solving posed problems from the context of the examples. The book assumes a reasonable proficiency in UML and at least one programming language (C++ is used

---

[3] There are always even more *bad* solutions to any problem as well!

in this book, but C, Java, Ada, or any equivalent 3GL can be used). This book is meant as a companion to the author's *Real-Time UML, Third Edition: Advances in the UML for Real-Time Systems,* but that book is not required to use this workbook. Other books that may help the reader by providing more depth in various topic areas include the author's *Doing Hard Time: Developing Systems with UML, Objects, Frameworks and Patterns,* Addison-Wesley, 1999, and *Real-Time Design Patterns: Robust Scalable Architecture for Real-Time Systems,* Addison-Wesley, 2002.

Goals

The goal of this book is simply to give the reader practical experience in modeling different aspects of real-time and embedded systems. While there is some introduction early in the first two chapters of the book and at the beginning of the problem chapters, the goal is to pose problems for the reader to model. The book includes a demo version of the powerful Rhapsody™ tool, but the use of that tool is not required to go through the exercises.

By the time you've reached the end of the book, you will hopefully have the expertise to tackle the real modeling issues that you face in your life as a professional developer of real-time and embedded systems.

Where to Go After the Book

If you're interested in tools, training, or consulting, see *www.telelogic.com* or *www.ilogix.com.* The author teaches classes and consults worldwide on UML, MDA, DoDAF, architectural design, design patterns, requirements modeling, use cases, safety critical development with UML, behavioral modeling, the development process improvement, project management and scheduling, and quite a bit more. You can contact him for training or consulting services at *Bruce.Douglass@telelogic.com.* He also runs a (free) yahoo group forum at *http://groups.yahoo.com/group/ RT-UML*—come on down! The I-Logix website also has many white papers available for downloading on different topics that may be of interest.

Evaluate UML on ARM

On the CD-ROM with this book, you will find the Rhapsody UML tool from Telelogic (for further information, see *What's on the CD-ROM?*). Do you want to evaluate the UML on an ARM microcontroller using C? Willert Software Tools delivers an evaluation version for the UML for generating ARM software. You can download this at *http://www.willert.de/rxf_eval.*

The evaluation software contains the following:

- Telelogic Rhapsody demo version

- Keil MicroVision/ARM compiler
  A good ARM compiler with a very usable IDE. Compiler, simulator and debugger integrated in a nice package.

- Willert WST52lt Bridge
  The Willert Framework RXF (modified version of the Rhapsody framework) optimized for small microcontrollers, code size under 4k ROM and 200 bytes of RAM, where speed and code size are critical.

The evaluation versions are limited but fully functional for the evaluation models. The generated models run in the Keil simulator. If you want to see the models run on real hardware you can order an evaluation kit for an attractive price. This kit contains a Keil MCB2130 Board with the Philips ARM7 chip and a uLink USB/JTAG debugger interface.

Bruce Powel Douglass, Ph.D.
Summer, 2006

Acknowledgments

I want to thank my editor, Tiffany Gasbarrini, for nagging me when I clearly needed it and being supportive when I need that too. My reviewers—Ian Macafee, Markus Rauber, Mark Richardson, Bart Jenkins, Frank Braun, and David McKean—were also helpful in keeping me honest. I have no doubt that errors remain, and I claim full responsibility for those. Most of all, I want to thank my family for keeping me sane, or at least trying to, while I burned the midnight electrons, creating this book.

About the Author

Bruce was raised by wolves in the Oregon wilderness. He taught himself to read at age 3 and calculus before age 12. He dropped out of school when he was 14 and traveled around the U.S. for a few years before entering the University of Oregon as a mathematics major. He eventually received his M.S. in exercise physiology from the University of Oregon and his Ph.D. in neurophysiology from the University of South Dakota Medical School, where he developed a branch of mathematics called autocorrelative factor analysis for studying information processing in multicellular biological neural systems.

Bruce has worked as a software developer in real-time systems for over 25 years and is a well-known speaker, author, and consultant in the area of real-time embedded systems. He is on the Advisory Board of the *Embedded Systems* and *UML World* conferences, where he has taught courses in software estimation and scheduling, project management, object-oriented analysis and design, communications protocols, finite state machines, design patterns, and safety-critical systems design. He develops and teaches courses and consults in real-time object-oriented analysis and design and project management and has done so for many years. He has authored articles for many journals and periodicals, especially in the real-time domain.

He is the Chief Evangelist[1] for Telelogic (formerly I-Logix), a leading producer of tools for real-time systems development. Bruce worked with Rational and the other UML partners on the specification of the UML. He is one of the co-chairs of the Object Management Group's Real-Time Analysis and Design Working Group. He is the author of several other books on software, including *Doing Hard Time: Developing Real-Time Systems with UML, Objects, Frameworks and Patterns* (Addison-Wesley, 1999), *Real-Time Design Patterns: Robust Scalable Architecture for Real-Time*

---

[1] Being a Chief Evangelist is much like being a Chief Scientist, except for the burning bushes and stone tablets.

Systems (Addison-Wesley, 2002), *Real-Time UML, Third Edition: Advances in the UML for Real-Time Systems* (Addison-Wesley, 2004), and several others, including a short textbook on table tennis.

Bruce enjoys classical music and has played classical guitar professionally. He has competed in several sports, including table tennis, bicycle racing, running, and full-contact Tae Kwon Do, although he currently only fights inanimate objects that don't hit back.

Bruce does extensive consulting and training throughout the world. If you're interested, contact him at *Bruce.Douglass@telelogic.com.*

What's on the CD-ROM?

- Demo version of the Rhapsody UML tool from Telelogic and various models of the solutions.
- README.TXT file in the root directory of the CD-ROM contains detailed instructions for how to install the software and obtain a temporary license to use the tool while working through the exercises in the book.

Introduction

What you will learn:

- **Basic modeling concepts of the UML**
  Overview of the UML. What's a design pattern?

- **Class and object models**
  What classes and objects are. How classes and objects work together in collaborations. Collaborations are the realizations of use cases. Includes packaging of logical elements.

- **Component and deployment models**
  Representing run-time artifacts and localizing them on processor nodes.

- **State machines and behavioral models**
  What are state machines and how do they model behavior?

- **Use case and requirements models**
  Capturing black-box behavior without revealing internal structure

Basic Modeling Concepts of the UML

The Unified Modeling Language (UML) is a third-generation object-modeling language standard, owned by the Object Management Group (OMG). The initial version of the OMG UML standard, 1.1, was released in November of 1997. Since then, a number of minor revisions and one major revision have been made. As of this writing, the current version of the standard is 2.0 and is available from the OMG at *www.omg.org*.

The UML is a rich language for modeling both software and systems, and it is the *de facto* standard for software modeling. There are a number of reasons for this, and it is the totality of all of them which, I believe, accounts for the phenomenal success of the UML. The UML is, first of all, relatively easy to learn and, once learned, relatively intuitive. Secondly, the UML is well-defined, and models written in the UML can be verifiable (if care is taken to be precise), so not only can the models be directly executed (with appropriate tools, such as Rhapsody™) but production-quality code can be generated from them. Third, there is *great* tool support; there are many vendors, and they have distinguished themselves in the market by emphasizing different aspects of modeling and development.

The notation used by the UML is graphical in nature, easy to master and, for the most part, simple to understand.[1] Although some people claim that the UML has *too many diagrams,* in reality there are only four basic types (see Figure 1.1). *Structural diagrams* include class, structure, object, package, component, and deployment diagrams. These are all basically differentiated, not on the contents of the diagram, but on their *purpose. Functional diagrams* emphasize functionality but not structure or behavior; functional diagrams include use case and information flow diagrams. *Interaction diagrams* focus on how elements collaborate together over time to achieve functional goals; interaction diagrams include sequence, communication (formerly known as "collaboration"), and timing diagrams. Lastly, *behavioral diagrams* focus on specification of behavior of individual elements; these include state machine and activity diagrams. Although the breadth of the notation can be a bit overwhelming to newcomers, in reality, complex systems can be easily developed with three core diagrams—class diagrams, statecharts, and sequence diagrams. The other diagrams

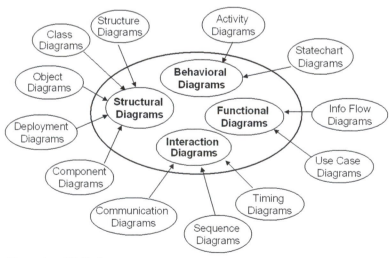

Figure 1.1 UML diagram types

---

[1]   Although, like all languages, it has subtle points as well.

can be used to model additional aspects of the system (such as capturing requirements, or how the software maps onto the underlying hardware). The additional
diagrams add value, certainly, but you only need the three basic diagram types to
develop systems and software.

The UML has a well-defined underlying semantic model, called the UML
metamodel. This semantic model is both broad (covering most of the aspects necessary for the specification and design of systems) and deep (meaning that it is possible
to create models that are both *precise and executable* and can be used to generate
source-level code for compilation). The upshot is that the developer can fairly easily
model any aspect of the system that he or she needs to understand and represent.
Figure 1.2 shows a screen shot of a model execution. Rhapsody uses color-coding
to depict aspects such as the current state, but color-coding doesn't show up well in
a black-and-white image. You can see the execution controls in the figure for step-
into, step-over, setting breakpoints, inserting events and so on. Rhapsody can also
dynamically create sequence diagrams that show the history of the interaction of
specified objects as they run.[2]

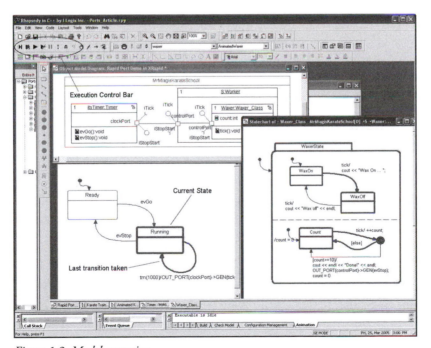

Figure 1.2 Model execution

---

2   The models in this book were created in Rhapsody. The accompanying CD contains a demo version
     of the tool for use in doing the exercises that form the bulk of this book. For more details, see the
     README.TXT file on the CD and see the tutorials available in the Help → List of Books menu.

The UML is a *standard*, unlike most modeling languages that are both proprietary and single-sourced. Using a standard modeling language means that the developer can select both tools and services from many different sources. For example, there are at least a couple of dozen different UML modeling tools. Since the tools focus on different aspects of modeling and have different price points, developers can find and select tools that best meet their own and their project's needs. For example, Rhapsody from Telelogic emphasizes the deep semantics of the UML, allowing the validation and testing of the user's models via execution and debugging using the UML notation. This execution can take place on the host development machine or on the final target hardware, and the generated code can then be used in the final delivered system. Other tools emphasize other aspects, such as drawing the diagrams but permitting more flexibility for a lower price point. The availability of so many different tools in the market gives the developer a great deal of latitude in tool selection. It also encourages innovation and improvement in the tools themselves. Because the UML is such a well-adopted standard, many companies provide training in the use and application of the UML. I spend a great deal of my time training and consulting all over the world, focusing on using UML in embedded real-time systems and software development.[3]

The UML is *applicable* to the development of software and systems in many different application domains. By now, the UML has been used in the development of virtually every kind of software-intensive system from inventory systems to flight control software. Being a maintained standard, the standard itself evolves over time, repairing defects, adopting good ideas and discarding ones that didn't pan out. The UML is used today to model and build systems that vary in scope from simple one- or two-person projects up to ones employing literally thousands of developers. The UML supports *all* the things necessary to model timeliness and resource management that characterize real-time and embedded systems. That means that developers need not leave the UML to design the different aspects of their system, regardless of how complex or arcane those things might be.

In this chapter, we'll introduce the basics of the UML. *This is intended more as a refresher than a tutorial.* Those desiring a more in-depth treatment of UML itself should pick up the companion book to this volume, *Real-Time UML, Third Edition: Advances in the UML for Real-Time Systems,* Addison-Wesley, 2004 by Bruce Powel Douglass. Additionally, there are many white papers available on the I-Logix website, *www.ilogix.com.*

---

[3]   Interested readers looking for professional services can contact me at *bruce.douglass@telelogic.com.*

Structural Elements and Diagrams

The UML has a rather rich set of structural elements and provides diagrammatic views for related sets of them. Rather arbitrarily, we'll discuss modeling small, simple elements first, and then discuss large-scale elements and model organization afterwards.

Small Things: Objects, Classes, and Interfaces

There are a number of elementary structural concepts in the UML that show up in user models: object, class, data type, and interface. These structural elements form the basis of the structural design of the user model. In its simplest form, an *object* is a data structure that also provides services that act on that data. An object only exists at run-time; that is, while the system is executing, an object may occupy some location in memory at some specific time. The data known to an object are stored in *attributes*—simple, primitive variables local to that object. The services that act on that data are called *methods*; these are the services invoked by clients of that object (typically other objects) or by other methods existing within the object. State machines and activity diagrams may enforce specific sequencing of these services when pre- and post conditions must be met.

A class is the specification of a set of objects that share a common structure and behavior. Objects are said to be *instances* of the class. A class may have many instances in the system during run-time but an object is an instance of only a single class. A class may also specify a statechart that coordinates and manages the execution of its primitive behaviors (called *actions*, which are often invocations of the methods defined in the class) into allowable sets of sequences.

If you look in Figure 1.3, you can see a basic class diagram. Note that objects are not shown. That is because when you show objects you are showing a snapshot of a running system at an instant of time. Class diagrams represent the set of possible object structures. Most often, when you create structural views, you will be more interested in creating class, rather than object, diagrams.

The DeliveryController object in the figure is an example class. It contains attributes, such as commandedConcentration (which is of type double) and the selected agent (i.e., drug) type. It provides methods, such as the ability to select an agent, to get the amount of drug remaining, and to set the commanded drug concentration. The DeliveryController class is shown as a standard box with three segments. The top segment holds the name of the class. The middle segment holds a list of the attributes. Note, however, that this list need not be complete—not all of the attributes owned by the class must be shown. The bottom compartment shows the methods of invocable services provided by the class.

Figure 1.3 shows other classes as well, and lines (called an *relations*—more on that later) connecting them. Some of these are shown in a simple, nonsegmented box, such as the TextView class. The attributes and methods, collectively known as *features* of the class, need not be shown. They are contained within the model and easily visible in the tool browser view, and may be exposed on the diagram if desired. But you have control over which features are shown on which diagrams.

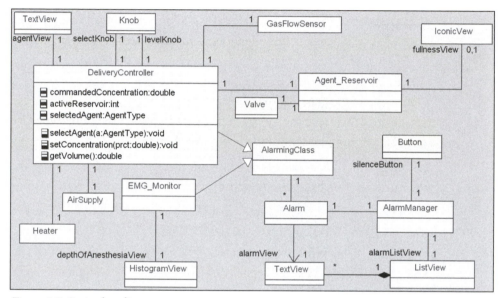

Figure 1.3 Basic class diagram

An *interface* is a named collection of services. Services come in two basic flavors. Operations are synchronous services that are invoked, or *called*, by clients. Event receptions are asynchronous services that are invoked by sending an asynchronous signal to the object accepting that event. An interface can contain either or both operations and event signals. Both operations and event signals can contain data, called parameters, passed with the invocation. The set of parameters, along with the name of the service, is called the *signature* of the service. A class that is compliant with an interface provides a method for every operation and an event reception for every signal specified in the interface.

Interfaces allow you to separate the specification of services that may be called on a class from the implementation of those services. A class defines *methods* (for operations) and *event receptions* (for signals, specified on the state machine). Both these methods and event receptions include the lines of code that implement the service. An operation or signal is a specification only and does not include such implementation detail.

Interfaces may not have implementation (either attributes or methods) and are not directly instantiable. A class is said to *realize* an interface if it provides a method for every operation specified in the interface, and those methods have the same names, parameters, return values, preconditions and postconditions of the corresponding operations in the interface.

Interfaces may be shown in two forms. One looks like a class except for the key word interface placed inside guillemots, as in «interface». This form, called a *stereotype* in UML, is used when you want to show the service details of the interface. The other form, commonly referred to as the "lollipop" notation, is a small named circle on the side of the class. Both forms are shown in Figure 1.4. When the lollipop is used, only the name of the interface is apparent. When the stereotyped form is used, a list of operations of the interface may be shown.

In Figure 1.4, two interfaces are shown, both of which happen to, in this case, only provide asynchronous event signals. The generalization arrow indicates the class that realizes the interface—that is, provides the services. Also shown are ports on the classes (more on ports later). The port is typed by the interfaces that are

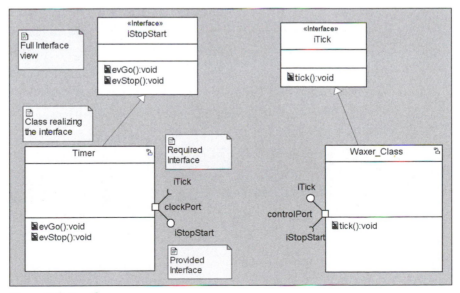

Figure 1.4 Interfaces

either offered or required across that port. In this case, the clockPort, a feature of the Timer class, offers the iStopStart interface services and requires the iTick interface. That means that to connect to that port, the other class must provide the iTick interface and require the iStopStart interface. The lollipop notation shows the provided (also known as "offered") interface while the socket notation indicates the required interface. This kind of modeling allows the developer to specify contracts that enable compatible elements to be linked together and is important, especially in architectural and component design.

In summary, an object is one of many possible *instances* of a class. A class has two notable features—attributes (which store data values) and methods (which provide services to clients of the class). Interfaces are named collections of operations that are *realized* by classes. Interfaces need not be explicitly modeled. Many useful systems have been designed solely with classes, but there are times when the addition level of abstraction is useful, particularly when more than a single implementation of an interface will be provided.

Structured Classes

One of the significant extensions in UML 2.0 is the notion of a *structured class*. A structured class is a class that is composed of *parts* (object roles) that are themselves typed by classes. Structured classes still have all the features and properties (such as attributes and methods) as normal classes, but they themselves internally own and orchestrate their internal parts. This is a huge advancement in the UML standard[4] because it allows the explicit specification of architectural containment structures. For example, systems contain subsystems as parts; these subsystems can contain sub-subsystems, and so on down to primitive (nonstructured) parts. In addition, since the common means for representing concurrency units (e.g., threads or tasks) is with active objects, then these active objects can contain the passive (nonconcurrent) objects as parts, allowing the clear specification of the concurrency architecture.

Figure 1.5 shows a typical structured class called SensoryAssembly. This class contains internal parts, such as a positioner part (typed by class Gimbal), a Mission-Tasker part (its class isn't exposed on the diagram) and a theSensor part (typed by class Sensor). Some of these parts are themselves structured classes and contain subparts. In this way, we can construct arbitrarily complex containment hierarchies.

---

4   Although Rhapsody has been doing this since its initial release.

A couple of things should be noted in Figure 1.5. First of all, the relation between a class and its parts is not a relation between classes—it definitely implies a composition relation between the classes, but the parts are not themselves classes. The class of the part might define different parts in different structured classes as well. Neither are parts objects. Every SensorAssembly has this same structure, so it's not referring to a specific singular object for its positioner. When we actually create an instance of the SensorAssembly, a specific singular instance of class Gimbal will play the role that this part specifies. But the part is not itself an instance. It is really a role that some instance will play when the system runs, and is typed by a class specification. The UML provides the term *part* for exactly this kind of object role.

The second thing to note is the presence of small squares on various class or part boxes. These small boxes are called ports and are discussed next.

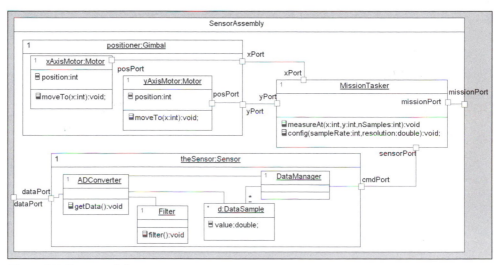

Figure 1.5 Structured classes

Ports

<mark>Classes may also have features called *ports*.</mark> Ports are named connection points of classes and, as mentioned before, are typed by the services offered and required across them.

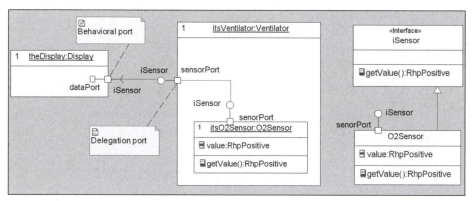

Figure 1.6 Ports

Relations

To achieve system-wide behavior, many objects must work together, and to work together, they must relate in some way. As part of structural modeling, the UML defines different kinds of relations among structural elements. We will focus primarily on the relations between classes, but these design-time relations between specifications have well-defined relations between instances of those classes.

The UML defines a number of different kinds of relations between classes. The most important of these are: association, generalization, and dependency.

Associations

<mark>The most basic relation is called the *association*. An association is a design-time relation between classes that specifies that, at run-time, instances of those classes may have a *link* (navigable relation between objects through which services may be invoked).</mark> The UML identifies three distinct kinds of associations: association, aggregation, and composition.

Association

An association between classes means simply that, at some time during the execution of the system, instances of those classes may have a link that enables them to invoke services from each other. Nothing is stated about *how* that is accomplished,

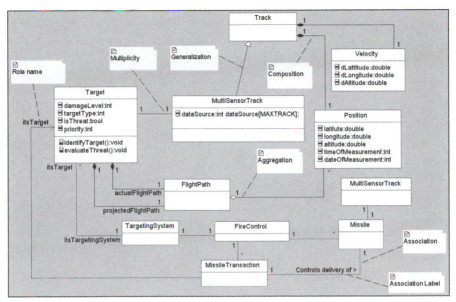

Figure 1.7 Class relations

or even whether it is a synchronous method call (although this is most common) or an asynchronous event signal. Associations are the specification of conduits that allow objects at run-time to send messages to each other. Associations are shown as lines connecting classes on class diagrams.

A number of aspects of an association between two classes can be specified. For example, the ends of the associations may have *role names*. These name the instances with respect to the other class. The most common implementation of associations is a pointer, and it is common to name the pointer with the role name at the opposite end of the association. For example, in Figure 1.7, the *Target* class might contain a pointer named *itsTargetingSystem* that would be dereferenced at run-time, just as the *TargetingSystem* class might contain a pointer named *itsTarget* that allows the TargetingSystem to send messages to the Target. I use the term "might" because that is a common implementation option, but there are other ways to implement associations as well.

Although somewhat less common, *association labels* may also be used, such as between the *MissileTransaction* and *Missile* classes. The label is normally used to help explain why the association exists between the two classes. In this case, the label "Controls delivery of >" indicates the purpose of the relation. The ">" is a common way to indicate the speaking perspective of the label; in this case, it is from the *MissileTransaction* to the *Missile*.

The *multiplicity* is probably the most important property of an association end. The multiplicity of an association end indicates the number of instances that can participate in the association role at run-time. This may:

- be a fixed number, such as "1" or "MAX_DATA_SOURCES"

- a comma separated list, such as "0,1" or "3,5,7"

- a range, such as "1..10"

- a combination of a list and a range, such as "1..10, 25", which means "one to ten, inclusive, or 25"

- an asterisk, which means "zero or more"

- or an asterisk with an endpoint, such as "1..*", which means "one or more."

In the figure, we see multiplicities on all the association ends with the exception of the directed association between the *Missile Transaction* and the *Target*. A normal line with no arrowheads means that the association is bidirectional; that is, an object at either end of the association may send a message to an object at the other end. If only one of the objects can send a message to the other and not vice versa, then we add an *open arrowhead* (we'll see later that the type of arrowhead matters) pointing in the direction of the message flow. The multiplicity at the nonnavigable end (the end without the arrowhead) is not normally specified. This is because the multiplicity at the "pointy end" of an association has influence on the implementation of the association in the class; you'll implement a 1-multiplicity with a pointer, but with * you might need an array of pointers. However, if the association end is non-navigable (as is the case in the "wrong" end of a directed association), then there *is* no implementation to change; hence its multiplicity is normally omitted.

All of these association adornments, except for perhaps multiplicity, are optional and may be added as desired to further clarify the relation between the respective classes.

An association between classes means that, at some point during the lifecycle of instances of the associated classes, there *may be* a link that enables their instances to exchange messages. Nothing is stated or implied about which of these objects comes into existence first, which other object creates them, or how the link is formed. Composition, the strong form of aggregation, *does* allocate create and destruction responsibilities, as we shall see in a moment.

Aggregation

An aggregation is a specialized kind of association that indicates a "whole-part" relation exists between the two objects. The "whole" end is marked with a white diamond, as in Figure 1.7. For example, consider the classes *FlightPath* and *Position*. The *FlightPath* class is clearly a "whole" that aggregates possibly many *Position* elements. The diamond on the aggregation relation shows that the *FlightPath* is the "whole." The "*" on the association end indicates that the list may contain zero or more Position elements. If we desired to constrain this to be no more than 100 waypoints, we could have made the multiplicity "0..100".

Since aggregation is a specialized form of association, all of the properties and adornments that apply to associations also apply to aggregations, including navigation, multiplicity, role names, and association labels.

Aggregation is a relatively weak form of "whole-part," as we'll see in a moment. No statement is made about lifecycle dependency or creation/destruction responsibility. Indeed, aggregation is normally treated in design and implementation identically to association. Nevertheless, it can be useful to aid in understanding the model structure and the relations among the conceptual elements from the problem domain. Sometimes it may not be clear which is appropriate, association or aggregation. In such cases, don't worry about it. Aggregation is typically implemented in exactly the same way as association, so it doesn't change the implementation. If you can't decide, pick one and change it downstream if necessary.

Composition

Composition is a strong form of aggregation in which the "whole" (also known as the "composite") has the explicit responsibility for the creation and destruction of the part objects. We've seen the structured class notation in the previous section between a class and its parts. Composition is the relation between the structured class and the class of those parts. There is a crucial distinction between aggregation and composition; in composition, you explicitly assign creation and destruction responsibility to the composite. For this reason, the multiplicity at the whole end of a composition relation is *always* 1. While an object can have multiple aggregation owners, it can only have at most one composite owner.

Because of object lifecycle responsibility, the composite exists before the parts come into existence and it exists after they are destroyed. If the parts have a fixed multiplicity with respect to the composite, then it is common to create those parts in its constructor (a special operation that creates the object) and destroy them in its destructor. With nonfixed multiplicities (e.g., "*"), the composite dynamically creates and destroys the part objects during its execution. Because the composite has creation and destruction responsibility, each part object can only be owned by a *single* composite object, although the part objects may participate in other association and aggregation relations. Composition is also a kind of association, so it can likewise have all of the adornments available to ordinary associations.

Figure 1.7 shows the filled diamond form, while Figure 1.5 shows the nested part (structured class) form. As a quick exercise-to-the-reader, redraw Figure 1.5 to use filled diamonds and redraw Figure 1.7 to use the nested part form. To be semantically correct, the filled diamond must be a relation between *classes* only and the nested part form is a relation between a class and parts (object roles).

Generalization

The generalization relation in the UML means that one class defines a set of features that are either specialized or extended in another. Generalization can be thought of as "is a type of" relation and therefore only has a design-time impact, rather than a run-time impact.

Generalization has many uses in class models. First, generalization is used as a means to ensure interface compliance, much in the same way that interfaces are used. Indeed, it is the most common way to implement interfaces in languages that do not have interfaces as a native concept, such as in C++. Also, generalization can simplify your class models because a set of features common to a number of classes can be abstracted together into a single superclass, rather than redefine the same structure independently in many different classes. In addition, generalization allows for different realizations to be used interchangeably; for example, one realization subclass might optimize worst-case performance while another optimizes memory size, while yet another optimizes reliability because of internal redundancy. Different subclasses can be substituted for an instance of the superclass and the collaboration will still "work."

Generalization in the UML means two things—inheritance and substitutability. First, it means *inheritance*—that subclasses have (at least) the same attributes, operations, methods, and relations as the superclasses they specialize. Of course, if the subclasses were *identical* with their superclasses, that would be unexciting, so subclasses can differ from their superclasses in either or both of two ways, specialization or extension.

Subclasses can *specialize* operations or state machines of their superclasses. Specializing means that the same operation (or action list on the statechart) is implemented differently from in the superclass. This is commonly called *polymorphism*. In order

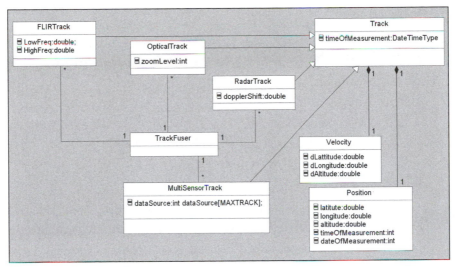

Figure 1.8 Class generalization

to make this work, when a class has an association with another that defines sub-classes, at runtime an instance of the first can invoke an operation declared in the second, and if the link is actually to a subclass instance, the operation of the subclass is invoked rather than that of the superclass.

Figure 1.8 shows a simple example. The *Track* class has a timeOfMeasurement attribute that identifies when the track was measured, and has composition relations with both the *Velocity* and *Position* classes. This is the base class for technology-specific tracks, for FLIR (Forward-Looking Infrared), optical, and radar sensors. Each of these has technology-specific additional data. The *TrackFuser* integrates data from three sensors monitoring the same aircraft into a multisensor track. Because these subclasses inherit features from the base class, they also have compositions with *Velocity* and *Position* classes as well.

The second thing that generalization means is *substitutability*. This means that an instance of a subclass is freely substitutable wherever there is an instance of the superclass. In Figure 1.8, a *Track* may provide services (such as returning its position, velocity, or combat ID) to a client. Because *MultisensorTrack* is a subclass of *Track*, at run-time the client of the *Track* class may actually have a link to an instance of the *MultisensorTrack* class. The client neither knows nor cares, because *MultisensorTrack* inherits all the capabilities of its superclass.

Dependency

Association, in its various forms, and generalization are the key relations defined within the UML. Nevertheless, several more relations are still useful. They are put under the umbrella of *dependency*. The UML defines four different primary kinds of dependency—Abstraction, Binding, Usage, and Permission. Each of these may be further stereotyped. For example, «refine» and «realize» are both stereotypes of the Abstraction relationship and «friend» is a stereotype of Permission. All of these special forms of dependency are shown as a stereotyped dependency (dashed line with an open arrowhead).

Arguably, the most useful stereotypes of dependency are «bind», «usage», and «friend». Certainly, they are the most commonly seen, but there are others as well. The reader is referred to the standard[5] for the complete list of "official" stereotypes.

The «bind» stereotype binds a set of actual parameters to a formal parameter list. This is used to specify parameterized classes (templates in C++-speak or generics in Ada-speak). This is particularly important in patterns because patterns themselves are parameterized collaborations, and they are often defined in terms of parameterized classes.

A parameterized class is a class that is defined in terms of more primitive elements, which are referred to symbolically without the inclusion of the actual element that will be used. The symbolic name is called a *formal parameter* and the actual element, when bound, is called an *actual parameter*. In Figure 1.9, *NumericParameter* is a parameterized class whose attributes and methods are defined in terms of an unspecified type, called *BaseType*. If we bind an actual parameter, say *double*, to be used whenever *BaseType* is used, we now have a class (*NumericParameter_double*) from which we can create instances; a fully-bound parameterized class is called an *instantiable class,* because we now have enough information to create instances of it. The «bind» dependency binds the actual parameter, *double*, to the formal parameter, *BaseType*.

Parametric classes can be subclassed, as they are in the figure. A *SafeNumericParameter* stores the data twice, once normally and once bit-inverted. It's not yet an instantiable class because we haven't specified what to use for *BaseType* yet. We do that when we create the *SafeNumericParameter_double*.

The diagram also shows the use of the «friend» dependency. This means that the *SafetyMonitor* class has access to all the features of the *SafeNumericParameter* regardless of their stated visibility—private, protected, or public. It's like the old saying goes, "You only show your friends your private parts."

---

[5]   Available at *www.omg.org*

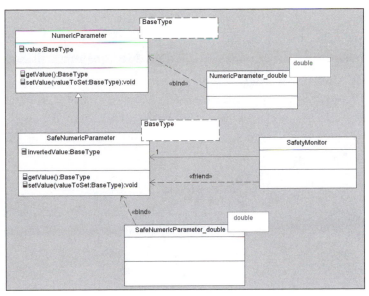

Figure 1.9 Dependency

Big Things: Subsystems, Components, and Packages

Classes, objects, and interfaces are, for the most part, little things. It takes collaborations of many of them to have system-wide behavior. In today's complex systems, it is important to think about larger-scale structures. The UML does provide a number of concepts to manage systems in the large scale, although most of the literature has not effectively explained or demonstrated the use of these features. And, to be honest, the UML specification does not explain them and how they interrelate very well either.

Packages are one such element. Packages are used *exclusively* to organize UML models. They are not instantiable and do not appear in any way in the running systems. Packages are model elements that can contain other model elements, including other packages. Packages are used to subdivide models to permit teams of developers to manipulate and work effectively together. They define a namespace for the model elements that they contain, but no other semantics. They are the primary model organizational units for configuration management—that is, in general, packages are the primary CIs (configuration items). When you check in or out a package from CM, you check in all the elements contained within that package. Rhapsody allows you to change the level of a CI down to the level of the individual class, but most developers find that too much detail to find all the relevant items to check out.

A package normally contains elements that exist only at design-time—classes and data types—but may also contain use cases and diagrams, such as sequence and class diagrams. These design pieces are then used to construct collaborations that realize system-wide functionality. Figure 1.10 shows that packages are drawn to look like a tabbed folder and may optionally show the elements that they semantically contain. It shows a single class within most packages, but they usually hold dozens or more elements. The dependency relations are optional but can be used to show compilation dependency.

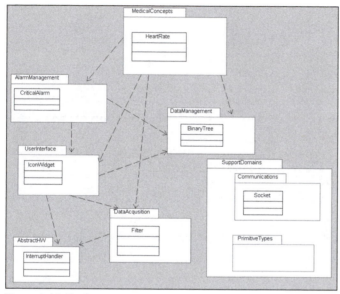

Figure 1.10 Packages

Subsystems are nothing more or less than classes—structured classes, specifically—and they are large-scale architectural building blocks. Unlike packages, subsystems are instantiable, meaning that you can create an instance of the type that occupies memory at run-time. A subsystem is used to organize the run-time system consisting of instances; the criterion for inclusion in a subsystem is "common behavioral purpose." The real work of a subsystem is implemented by the parts contained within the subsystem; the subsystem offers up the collaborative behavior of those elements. Subsystems organize and orchestrate behavior of their internal parts, but the "real work" is done by what is sometimes called the *semantic objects* of the system—the primitive objects that actually perform the bottom-level functionality. A subsystem is at a higher level of abstraction of the system than these primitive semantic objects and this level of abstraction allows us to view and manipulate the structure and behavior of complex systems much more easily.

In UML 2.0, components are basically also structured classes that provide opaque language-independent interfaces. There are some subtle metamodeling distinctions between components and subsystems, but in use they are practically identical.

Behavioral Elements and Diagrams

What we've discussed so far is the definition of structural elements of the system: classes and objects (in the small) and systems, subsystems, and components (in the large). As developers, we are usually even more concerned about how these structural elements behave dynamically as the system runs. Behavior can be divided up into two distinct perspectives—how structural elements act in isolation and how they act in collaboration.

In the UML metamodel, *ModelElements* are the primary structural elements that have behavior. *Classifiers* (which are types of *ModelElements*) also have *BehavioralFeatures*, specifically *Operations*, and the realization of operations, *Methods*. In practice, we are primarily concerned with the specification of the reactive behavior of only certain *Classifiers* (classes, objects, subsystems, components, and use cases) and certain other *ModelElements (Actions, Operations,* and *Methods).*

Actions and Activities

An action is a "specification of an executable statement that forms an abstraction of a computational procedure that results in a change in the state of the model, and can be realized by sending a message to an object or modifying a link or a value of an attribute." That is, it is a primitive thing, similar in scope to a single statement in a standard source-level language, such as "++X" or "a=b+sin(c*PI)". The UML identifies a number of different kinds of actions, such as creating an instance, invoking a method, or generating an event.

While the UML specifies what is called *action semantics,* it doesn't specify a language syntax to use. Some tools use proprietary abstract action languages that are converted to a target language during the model compilation process. This has the (claimed[6]) advantage of portability of the application from one source language to another. The downside of the approach is that debugging at the code level is usually very difficult and if you—ever—modify the source code, then the connection between the model and the code is irretrievably broken. By far, most developers prefer to specify actions in their intended source language, whether it is C, C++,

---

[6] I say "claimed" because, in practice, the generation of a concrete language with these tools is not as simple and obvious as it could be, and most of the tools that work this way produce some really ugly and inefficient code for the target language.

Java, Ada, or something else. For one thing, *they already know* that language. For another, these languages have standards (as opposed to proprietary action languages) and multiple compilers to support them on many different platforms. Debugging is far simpler, and bringing code changes back into the model can be relatively easy, depending on the tool.

Actions have "run-to-completion" semantics, meaning that once an action is started it will run until it is done. This does not mean that an action cannot be preempted by another action running in a higher-priority thread, merely that when the context executing that action returns from preemption, it will continue executing that action until it is complete. This means that if an object is executing an action, that action will run to completion even if that object receives events directing it to do something else. The object will not accept the incoming events until the action has completed.

An *activity* is an action that runs when a Classifier is in a state and is terminated either when it is complete or when the Classifier changes state. That is, activities do not have run-to-completion semantics. An object executing an activity may receive an event that triggers a transition, exiting the state and terminating the activity. Thus, the UML allows the modeling, at a primitive level, of both interruptible and noninterruptible behaviors.

Operations and Methods

An operation is a specification of an invocable behavior of a Classifier, whereas a method is the implementation of an operation. That is, an operation is a specification of a method. Operations are synchronously invoked and are logically associated with CallEvents in the UML metamodel. Operations have typed parameter lists, as you might expect, and can return typed values. It is common to use an operation call as an action on a state behavior.

Modeling of the behavior of a method is done primarily in two ways. First and most common is to simply list, in the selected action language, all the actions to implement the method. This is clearly the best approach for simple methods. The second approach, which will be described shortly, is to model the operation with an activity diagram.

Activity Diagrams

In UML 1.x, activity diagrams were nothing more (or less) than a state machine with a different syntax. That is, both activity diagrams and state machines had a common underlying behavioral metamodel. In UML 2.0, however, that's been changed.

In UML 2.0, activity diagrams execute based on *token execution semantics*; that is, when an activity receives the run token from the activity that precedes it, it can run. When it is done executing, it passes the token on to the next activity in the sequence. The transitions from activity to activity (or action to action) are taken when the predecessor activity completes—no events are used. In Figure 1.11 the rounded rectangles are actions or activities. The arrows indicate the flow transitions. The transition with a ball on one end is called the *initial pseudostate* and indicates where execution begins when the behavior starts.

Activity diagrams contain operators. The diamonds indicate selection based on the execution of guards (shown in square brackets). These guards are Boolean expressions with no side effects. Transition is taken when the guard evaluates to true. The *else* transition is taken only if all other transitions are false. The bar with a single input transition and multiple exiting transitions is called a fork. It differs from selection in that, with selection, at most one outgoing transition is taken; with a fork, *all* outgoing transitions are taken. That is, a fork indicates the presence of logical threads of execution running simultaneously.[7] A join, a bar with multiple incoming transitions, collapses multiple logical threads into one. The termination state indicates the end of the behavior.

Actions and activities can be owned by various objects; they can invoke methods from other objects to which they have links. The allocation of the actions and activities is most commonly done with *swimlanes*. In the figure, swimlanes are the named rectangles containing the actions and activities. The name of each swimlane identifies a class that provides the invoked method.

Another new thing added in UML 2.0 is the notion of pins. Just as an activity (or action) corresponds to a function, a pin corresponds to a parameter of a function. Thus, pins can be shown as either the origin of transitions (indicating the passing of the data to the parameter specified by the pin) or its termination (indicating the reception of data to the parameter). Pins at the origin of a transition are called *output pins* and pins at the termination of a transition are called *input pins*. Input and output pins are indicated in Figure 1.11. Data can also be indicated with *object in state*, a notational hangover from UML 1.4. For example, "validatedFlightPlan" is such an element; its state or condition is shown in square brackets inside the box. In UML 2.0, it is more common to use pins.

---

[7]   I say "logical" because they don't have to be implemented with OS threads. They can be executed sequentially but they are logically concurrent so the developer cannot know which is executed first. That is, they are execution-order independent with respect to each other.

Activity diagrams are like flowcharts on steroids. The most common use for activity diagrams is to model algorithms. As such, their most common applications are to represent the behavior of a method of a class or the behavior of an algorithmic use case.

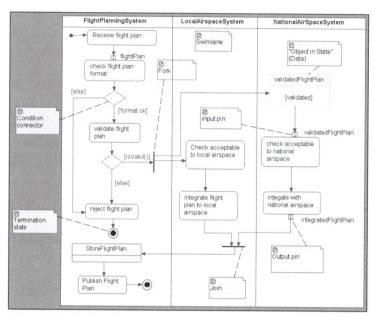

Figure 1.11 Activity diagram

Statecharts

A finite state machine (FSM) is a machine specified by a finite set of conditions of existence (called "states") and a likewise finite set of transitions among the states triggered by events. An FSM differs from an activity diagram in that the transitions are triggered by events (primarily) rather than being triggered when the work done in the previous state is complete. Statecharts are primarily used to model the behavior of *reactive* elements, such as classes and use cases, that wait in a state until an event of interest occurs. At that point, the event is processed, actions are performed, and the element transitions to a new state.

Actions, such as the invocation of an operation, may be specified to be executed when a state is entered or exited, or when a transition is taken. The order of execution of actions is exit actions of the predecessor state, followed by the transition actions, followed by the entry actions of the subsequent state.

The UML uses statecharts as their formal FSM representation, as they are significantly more expressive and scalable than "classical" Mealy-Moore FSMs. UML state machines, based on Dr. David Harel's Statechart semantics and notation,[8] have a number of extensions beyond Mealy-Moore state machines, including:

- Nested states for specifying hierarchical state membership

- And-states for specifying logical independence and concurrency

- Pseudostates for annotating commonly needed specific dynamic semantics.

Figure 1.12 shows some of the basic elements of a statechart for dialing a number with a telephone. It includes basic or-states and transitions, as well as a few less-elementary concepts, including nested states and initial pseudostates.

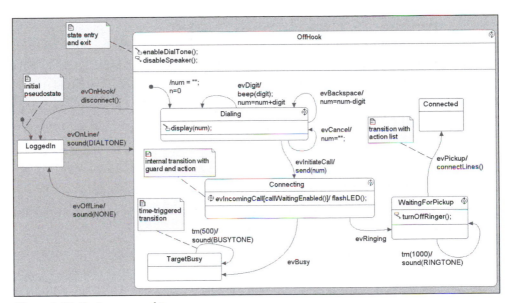

Figure 1.12: Basic state machine

Transitions are arrowed lines coming from a predecessor state and terminating on a subsequent state. Transitions usually have the optional *event signature* and *action list*. The basic form of an event signature is:

event-name '(' parameter-list ')' ['guard'] '/' action-list

---

8   Dr. David Harel, inventor of statecharts, was one of the founders of I-Logix (now a part of Telelogic) and was instrumental in the creation of the Statemate and Rhapsody tools.

The event-name is simply the logical name of the event class that can be sent to an instance of the Classifier at run-time, such as 'evOnHook' or 'tm' in Figure 1.12. The UML defines four distinct kinds of events that may be passed or handled:

- SignalEvent—an asynchronously sent event

- CallEvent—a synchronously sent event

- TimeEvent—an event due to the passage of an interval of time (most common) or arrival of an epoch

- ChangeEvent—a change in a state variable or attribute of the Classifier

Asynchronous event transfer is always implemented via queuing of the event until the element is ready to process it. That is, the sender "sends and forgets" the event and goes on about its business, ignorant of whether or not the event has been processed. Synchronous event transfer executes the state processing of the event in the thread of the sender, with the sender blocked from continuing until that state processing is complete. This is commonly implemented by invoking a class method called an event handler that executes the relevant part of the state machine, returning control to the sender only when the event processing is complete. Rhapsody refers to CallEvents as "triggered operations"; they are like normal operations in that they are synchronous, but don't have a standard method body. Their method body is the action list on the state machine.

Events may have parameters, which are typed values accepted by the state machine which may then be used in the guard and actions in the processing of the event. The statechart specifies the formal parameter list, while the object that sends the event must provide the necessary actual parameters to bind to the formal parameter list. Rhapsody uses a slightly peculiar syntax for passing events. Rhapsody automatically creates a struct named *params* that contains the named parameters for every event that carries data. If an event *e* has two parameters, x and y, to use these in a guard, for example, you would dereference the params struct to access their value. So a transition triggered by event *e* with a guard that specified that x must be greater than 0 and y must be less than or equal to 10 for the transition to be taken would look like:

$$e[params\text{->}x\text{>}0 \ \&\& \ params\text{->}y\text{<=}10]$$

Time events are almost always relative to the entry to a state. A common way to name such an event (and what we will use here) is 'tm(interval)', where 'interval' is the time interval parameter for the timeout event.[9] If the timeout occurs before

---

9   Another common way is to use the term "after(interval)".

another specified event occurs, then the transition triggered by the timeout event will be taken; if another event is sent to the object prior to the triggering of the timeout, then the time out is discarded. If the state is reentered, the timeout interval is started over from the beginning.

If a transition does not provide a named event trigger, then it is activated by the "completion" or "null" event. This event occurs either as soon as the state is entered (which includes the execution of entry actions for the state) or when the state activities complete.

A guard is a Boolean expression contained within square brackets that follows the event trigger. The guard should return only TRUE or FALSE and not have side effects. If a guard is specified for a transition and the event trigger (if any) occurs, then the transition will be taken if and only if the guard evaluates to TRUE. If the guard evaluates to FALSE, then the triggering event is quietly discarded and no actions are executed.

The action list for the transition is executed if and only if the transition is taken; that is, the named event is received by the object while it is in the predecessor state and the guard, if any, evaluates to TRUE. The entire set of exit actions—transition actions—entry actions is executed in that order and is executed using run-to-completion semantics, as noted previously. Figure 1.12 shows actions on a number of different transitions, as well as actions on entry for the Dialing state, on exit for the WaitingForPickup state, and both entry and exit for the OffHook state.

In addition to entry and exit actions, states may have reactions, also known as *internal transitions*. These are actions taken when the object is in the specified state and the triggering event is received, but in this case, the state is not changed. UML Comments in Figure 1.12 indicate some entry actions, exit actions, and internal transitions. Rhapsody uses small icons beside the action to indicate when the actions (on entry, exit, or reaction) inside the state are executed.

If you look at Figure 1.12, the rounded rectangles are states. These are called *or-states* because at any level of state machine abstraction, the object must be in only one of these states. In the figure, the Telephone object can either be LoggedIn or Offhook. These are or-states at the same level of abstraction. The initial pseudostate indicates which you start in when the object is created. Rhapsody provides an IS_IN(state) macro for all state machines; IS_IN(LoggedIn) will return TRUE if and only if the specified Telephone object is in the LoggedIn state.

The OffHook state is also a *composite state* and contains nested states, such as Dialing, Connecting, and WaitingForPickup. Within the OffHook state, its nested states are or-states; that is, if the object is in the OffHook state, then it must be in *exactly one* of the nested states. If the Telephone object is in the WaitingForPickup state, then not only does IS_IN(WaitingForPickup) return TRUE, so does IS_IN(OffHook). That is a basic characteristic of nested states; if you're in the bathroom, then yes, you ARE also in your house, and you can only be in one room of the house at a time and in one house at a time.

The evOnLine event enters the OffHook and therefore must enter a nested state as well, but which one? The answer is indicated with the initial pseudostate inside the OffHook state—it indicates which of those nested states is the default. The default can be bypassed by drawing a transition directly to a different nested state, if desired. Note that there are two transitions from the OffHook state back to the LoggedIn state. They come from the composite, so they apply whenever OffHook is the current state, regardless of which nested state in OffHook is currently valid. In Mealy-Moore state machines, which lack nesting, those two transitions would each have to come from every one of the nested states and go to the LoggedIn state to provide the same behavior.

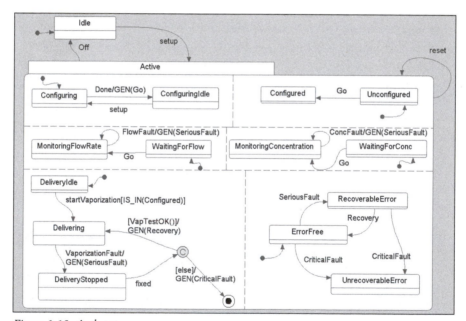

Figure 1.13 And-states

Figure 1.13 shows an important additional concept in statecharts, "and-states." While or-states are disjoint and exclusive, and-states are disjoint but not non-exclusive. These are equivalent to the logical threads in activity diagrams; they are usually not actually modeled as separate OS threads, but the order of execution between the and-states is not inherently knowable. Event receptions have object-wide scope. This means that if an object has multiple active and-states and receives an event, *all* active and-states receive a copy of the event and are free to act on that event or discard it as appropriate. Furthermore, the order in which those and-states process their copies of that event is not known. If you want to specify a particular sequencing, then use or-states rather than and-states. For idioms for synchronizing action execution across and-states, and for a more detailed discussion of UML state machines, refer to Chapter 3 of *Real-Time UML, Third Edition*[10] and Chapters 7 and 12 in *Doing Hard Time*.[11]

Interactions

The UML models the collaborative behavior of multiple entities working together as *interactions*. The UML provides three primary diagrams to represent interactions: communication diagrams (known in UML 1.x as "collaboration diagrams"), sequence diagrams, and timing diagrams. Communication diagrams are basically object diagrams with messages shown with numbers. Sequence diagrams show object roles as vertical lifelines with message sequences going down the page. Timing diagrams depict changes in state or value over linear time. Neither communication nor timing diagrams are widely used, so we will limit our discussion here to sequence diagrams.

Sequence diagrams depict object roles, which might represent objects, subsystems, systems, or even use cases, interacting over time. The vertical lines in Figure 1.14, called "lifelines," represent the object (or object role). Objects (or object roles) can both send and receive messages. Messages are shown by the arrowed lines going from one lifeline to another.

The sequence diagram shows an *exemplar* or "sample execution" of some portion of the system under specific conditions. Such an exemplar is commonly called a *scenario*. The messages may be synchronous (shown with a solid arrowhead) or asynchronous (shown with an open arrowhead). Sequence flows, more or less, from the top of the page downwards. Additionally, state or condition of the lifeline can be shown, as can constraints. In Figure 1.14, a timing constraint is shown on the right of the screen. A more general constraint is shown anchored to the alarm(Gas_Supply_Fault) limiting the types of faults that are reported using this value.

---

[10] Douglass, Bruce Powel, *Real-Time UML, Third Edition: Advances in the UML for Real-Time Systems*, Addison-Wesley, 2004.

[11] Douglass, Bruce Powel, *Doing Hard Time: Developing Real-Time Systems with UML, Objects, Frameworks, and Patterns*, Addison-Wesley, 1999.

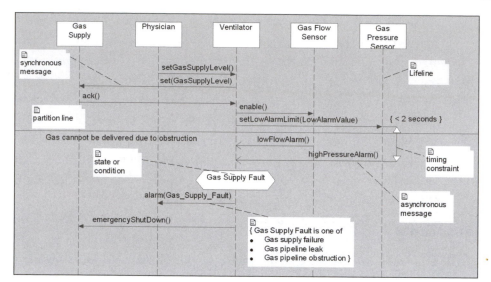

Figure 1.14 Sequence diagram

UML 2.0 significantly extends sequence diagrams over previous versions. Diagrams may contain, essentially, subdiagrams called *interaction fragments.* Each interaction fragment can have an operator, such as loop, opt (for "optional"), alt (for "alternative"), ref (for "reference"), par (for "parallel") and so on. These interaction fragments and operators greatly enhance the ability of sequence diagrams as specification tools. Figure 1.15 shows three interaction fragments. One has a parallel operator indicating that it contains regions that execute concurrently. Within that interaction fragment are nested two more, with loop operators indicating that they repeat until some termination condition is reached.

What I think is the most significant extension to sequence diagrams in UML 2.0 is the ability to formally decompose them. This can be done either "horizontally" by using an interaction fragment with the "ref" operator, or "vertically" by setting a reference from one lifeline to a separate sequence diagram that shows the same scenario at a more detailed level of abstraction. The next three figures show how these decomposition mechanisms can be used.

Figure 1.16 shows the "high-level" sequence diagram for an industrial robot system. The user sets up the system and, based on the task plan, the controller commands the robot to achieve the tasks. The robot itself has internal parts: two angular joints (called the knee and the elbow) and a rotating manipulator, which can grab and control tools. This high-level sequence diagram contains two references to more detailed interactions. The first of these is the "Setup System" referenced

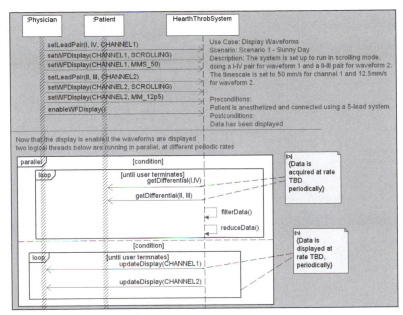

Figure 1.15 Interaction fragments

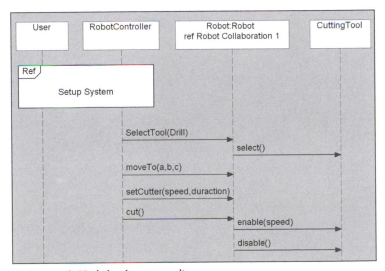

Figure 1.16 High-level sequence diagram

interaction fragment. If we open up that diagram (a right-click in Rhapsody), we see the details shown in Figure 1.17.

Even more valuable is the ability to decompose the lifeline. This mechanism allows the same scenario to be viewed at many different levels of abstraction without overwhelming the viewer by putting everything on a single, huge diagram. Figure 1.18

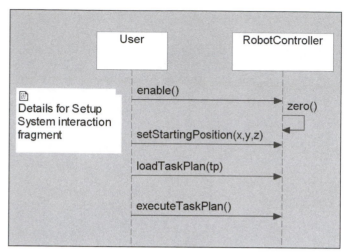

Figure 1.17 Referenced interaction fragment

shows the details of how the internals of the Robot interact to achieve their roles in this same scenario. The ENV lifeline is the connection between the high- and low-level interactions. At the high level, a message going to the Robot lifeline comes *out* of the ENV lifeline on the more detailed diagram. Conversely, a message going *into* the ENV lifeline on the more detailed diagram comes out of the Robot lifeline on the higher-level sequence diagram.

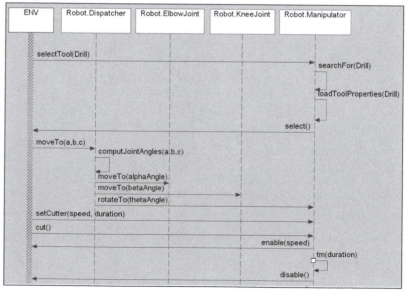

Figure 1.18 Decomposed lifeline

Use Case and Requirements Models

A use case is an explicitly named capability of a system or large-scale element of a system. It is a collection of operational and related quality of service requirements around a functional aspect of the system. It should return a result to one or more actors, and should not reveal or imply anything about the internal implementation of that functionality. In a medium-sized system, a use case might represent 8–20 pages of detailed requirements.

There are two different kinds of requirements typically applied against a system or system element: functional requirements and so-called nonfunctional or quality of service (QoS) requirements. Functional requirements refer to what the system needs to do, as in "The system shall deliver the anesthetic drug Halothane in inhalant form." QoS requirements refer to *how well* the functional aspects are to be achieved, as in "The system shall deliver the anesthetic drug Halothane in inhalant form *and maintain the commanded gas concentration within 0.5% by volume. From a drug-free gas concentration, a maximum of 10% by volume shall be achieved in no less than 10 minutes at a breathing rate of 10 breaths/minute and a tidal volume of 600 ml.*"

Use cases are drawn as ovals that associate with actors, indicating that the realizing collaborations interact in meaningful ways with those specified actors. In addition, use cases may relate to other use cases, although the novice modeler is cautioned against overuse of these relations.[12] The three relations among use cases are generalization (one use case is a more specialized form of another), includes (one use case includes another to achieve its functional purpose), and extends (one use case may optionally add functional aspects to another). An example of a use-case diagram for a medical anesthesia system is shown in Figure 1.19.

Since a use case provides little more than a name and relations to actors and other use cases, the detailed requirements must be captured somewhere. The Harmony process refers to this as "detailing the use case." There are two complementary approaches to detailing use cases—by example and by specification. In both cases, however, the internals of the structure of the system cannot be referred to, since the requirements should be captured in an implementation-free manner.

---

[12] In the author's experience, it is all too easy to use the use-case relations in a misguided attempt to functionally decompose the internals of the system, which is *not the point*. The purpose of use cases is to define the functional behavior of a system, subsystem, or other large-scale classifier in an implementation-free way.

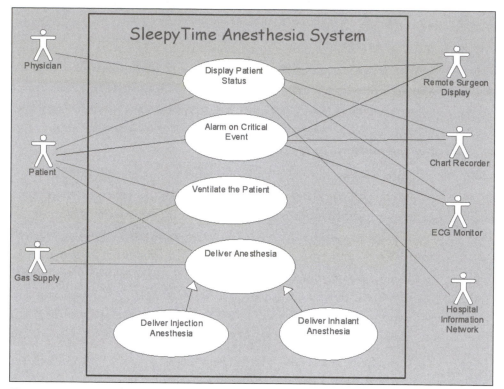

Figure 1.19 Use-case diagram

Use cases may be thought of as "bags" that contain related detailed requirements. These requirements, as mentioned above, may be a combination of functional and QoS requirements. These details are captured in one or both of two ways: scenario modeling with sequence diagrams or specification, especially with state machines.

Scenario modeling of use cases involves the creation of sequence diagrams that capture different scenarios of the use case. Each scenario captures a very specific system-actor interaction. These scenarios capture messages sent to the system from the actors and from the system to the actors, as well as the allowable set of sequences of such messages.

When doing scenario modeling, it is important to remember the purpose is to capture requirements, not to functionally decompose the internals of the system. Therefore, for requirements scenarios, the only classifiers that may appear are the system actors and the system or the use case, not pieces internal to the system. In the case of subsystem use cases, then peer subsystems are treated as actors to the subsystem of concern.

The other approach to requirements capture is via *specification*. The specification may be done in hundreds or thousands of textual statements but text is best used in conjunction with a more formal language, such as statecharts or activity charts, to define all possible scenarios. When statecharts are used in this way, messages from the actors are represented as events on the statechart while messages from the system to the actors are shown as actions on the statechart.

Summary

That's it—your brief refresher on UML! Classes and objects are the usual things that people first consider when they think about the UML but, clearly, the UML is more expansive and expressive than this. The UML also includes primitive behaviors—actions and activities—and ways to specify the allowable sequences of their execution with activity diagrams and state machines. The UML provides means for showing example interaction, primarily with sequence diagrams as well. Finally, use cases cluster requirements into usable, coherent functional units. This chapter was not meant to be your *only* introduction to the UML. Although we hit the highlights, we didn't go into much depth, and many useful features of the UML weren't discussed at all. As mentioned previously, this book is meant to be a companion volume to the author's *Real-Time UML, Third Edition*, and the interested reader is referred there for more detail on the UML.

The UML is a language and is intentionally process-agnostic. Any reasonable process can be used effectively with the UML. However, not all processes are equally effective. In the next chapter, we'll discuss one process, called the *Harmony process*, that has been developed by the author, in collaboration with others over the years, and applied very effectively in the development of real-time and embedded systems. The problems and answers that come later in this book will follow, more or less, the Harmony process, so the next chapter will provide a brief introduction.

Check Out the CD-ROM

This book contains a CD-ROM with various things on it, such as the models worked on in this book and a trial copy of Rhapsody. All of the examples in this book are modeled in Rhapsody. Check out the README.TXT file on the CD-ROM for instructions on how to install the tool as well as how to get the required license. You don't need to use Rhapsody to work on these problems, but it is recommended.

Rhapsody is a very powerful and highly capable tool. Once you've installed it, I *highly recommend* that, before you start trying to solve the problems presented in this book, you spend time going through the *entire* tutorial to learn how to "drive" the tool, where things are in the menu structure, how to do execution and debugging, and so on. To get to the tutorial, run Rhapsody, then select Help → List of Books. In that list, you will see Rhapsody in C++ Tutorial. Select that (or whichever language you want to use) and go through the entire tutorial. It will make performing the exercises much, much easier. Don't worry—I'll wait patiently until you get back! ☺

2

The Harmony Process

What you will learn:

- **The Harmony development process**
  - Why process?
  - Harmony process overview
    - Key enabling technologies
    - Process timescales
  - Harmony process variants
  - Harmony microcycle in detail
    - Party!
    - Analysis
    - Design
    - Translation
    - Test

Introduction

A methodology consists of a language to specify elements and relations of interest and a process that tells the developer what parts of the language to use, how to use them, and when to use them. The Harmony process[1] uses UML and variants, such as the Systems Modeling Language (SysML) or the Department of Defense Architecture Framework (DoDAF) UML profiles, as the language. The Harmony process also specifies an integrated set of workflows to guide the developer so that they can use the UML to its fullest advantage in developing robust, capable, and safe systems.

---

[1] The Harmony process is basically the next revision in the ROPES (Rapid Object-oriented Process for Embedded Systems) process, discussed in the author's previous books with greatly expanded coverage for systems engineering.

There is a very broad range of development processes in use today, from "We don't need no stinking process" to very formal rigorous processes. This chapter begins the discussion with an overview of the two major variants of the Harmony process, and then gives detailed workflows for each of the phases in the process.

Beginning with the next chapter, problems will be presented for you to work on. The set of problems is presented, roughly, in the order in which they appear when you follow the Harmony process. The order of these problems may vary from how they would appear to you if you follow a different process.

The Harmony Development Process

A process is an integrated set of workflows. Each workflow takes some aspect, typically a phase in the process, and elaborates what activities are necessary for the workers to accomplish, when and how they are going to accomplish it, and what artifacts they generate.

A good process provides guidance on an effective way to develop high-reliability systems at minimal costs. Far too many processes are either completely under-specify workflows or waste valuable developer time and resources doing the wrong things, such as generating reams of paperwork. A good process usually produces some paper artifacts, but only those that add value, and even then in a cost-effective manner.

Why Process?

The basic reason why we, as software and system developers, should be concerned about and use a good process is to improve our lives and our products. Specifically, a good process:

- Provides a project template to guide workers through the development and delivery of a product
- Improves product quality in terms of
    - Decreased number of defects
    - Lowered severity of defects
    - Improved reusability
    - Improved stability and maintainability
- Improves project predictability in terms of
    - Total amount of effort
    - Length of calendar time required for completion
- Communicates project information appropriate to different stakeholders in ways that allow them to use it effectively

If you have a process that doesn't achieve these goals, then you have a bad process and should think about changing it for the better. These goals can be achieved with a good process or they can be inhibited by a bad process.

So, what's a process? In Harmony, we define a process to be:

> *A **process** is the specification of a sequenced set of activities performed by a collaborating set of workers resulting in a coherent set of project artifacts, one of which is the desired system.*

A process consists of worker roles, the "hats" worn by workers while doing various project activities. Each activity results in the creation or modification of one or more artifacts. For example, most processes have requirements capture (activity) somewhere early on before design occurs. This is performed by a requirements analyst (worker) acting as a software modeler (a worker role), and might result in an artifact, such as a portion of the software model from which code will be generated. Figure 2.1 depicts these fundamental aspects and relations inherent in a development process.

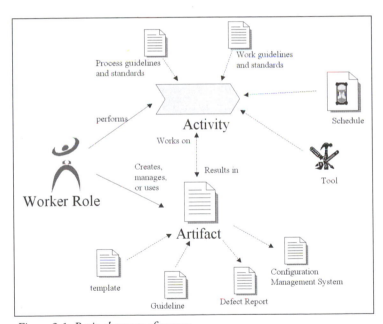

Figure 2.1 Basic elements of process

The activities are the tasks that the worker does in performance of his or her duty. The activities are grouped together into workflows focused around a common thread, such as the work done:

- in a development phase
- to achieve a specific goal

- to create a specific artifact

- by a particular worker role.

A process is normally organized into phases, which might be thought of as the largest-scale activities. Each phase is specified with one more workflows. Each workflow is a sequenced set of activities—simple tasks performed by workers—with resulting artifacts. A common way to represent workflows is with UML activity diagrams, and that approach will be followed here.

Artifacts are the things created or modified during activities. The singular most important artifact is The System being produced but there are many others, such as the source code, the software model, the requirements specification, test vectors, and so on. Generally speaking, every activity results in the creation or modification of at least one artifact.

The Harmony process, described in more detail in the next section, is applicable to (and in current use in) projects of widely different scale. Harmony achieves this scalability in a couple of different ways. First, the process is viewed at multiple timescales—macro, micro, and nano. Smaller projects will give much more attention to the micro and nano cycles, but as the projects grow in size, more attention is shifted to the macro scale to organize and orchestrate the entire development process. Secondly, a number of artifacts are optional and created during the process only as needed. Hazard analysis, for example, is only used for safety-critical applications. The subsystem architecture view, for another example, is only created when systems are large enough to profit from such decomposition. In fact, the Harmony process has two major variants: one, called the Full Harmony process, includes a detailed

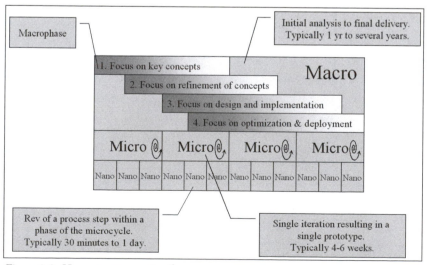

Figure 2.2 Harmony process timeframes

systems-engineering process that precedes the subsystem development, and another, called Harmony-SW, focuses on software development only.

Despite its well-known problems, waterfall lifecycle is still by far the most common way of scheduling and managing projects. Nevertheless, the most fundamental issue with the waterfall lifecycle is that defects introduced early in the process are not identified or fixed until late in the process. Certain kinds of strategic defects—requirements and architectural defects, specifically—are three or four orders of magnitude more expensive to repair in the waterfall lifecycle because they have broad sweeping implications. This is inherent in the waterfall lifecycle because testing comes at the end. The longer you wait to identify and repair defects, the more they have become entrenched and the greater the number of dependencies on the flawed aspects. Put another way, the problem with the waterfall lifecycle is that it fundamentally assumes each step in the process can be completed more-or-less without serious defects, but in fact that is demonstrably untrue. When the defects are finally identified and repaired, the cost is very high.

The spiral (also known as the iterative) lifecycle has become popular to address the concerns associated with the waterfall lifecycle. The basic advantage of the spiral lifecycle is that the system is tested far earlier and far more often. This results in the identification and repair of defects much earlier and at a significantly reduced cost. The spiral lifecycle essentially breaks up the development project into a set of smaller projects, and incrementally adds capabilities to the system, but not before validating the ones already present. Each addition of a set of capabilities is called a "spiral" or "increment." Each of these subprojects is more limited in scope, is produced with much greater ease, and has a much more targeted focus than the entire system. The result of each spiral is what Harmony calls an iterative prototype—a functional, high-quality system that may not be as complete (or perhaps not done in as high fidelity) as the final system. Nevertheless, the prototype does correctly implement and execute some portion of the requirements and/or reduce some set of risks and contains the actual code that will ship with the product, once complete.

The Harmony process can be conceptualized as occurring simultaneously in three different scales or time frames (see Figure 2.2). The macrocycle process occurs over the course of many months to years and guides the overall development from concept to final delivery. The Harmony macro process has four primary, but overlapping, phases. Each macrophase actually contains multiple microcycles,[2] as we will see shortly, and the result of each microcycle is the production of an iterative prototype.

---

[2]   That is, spirals.

The macrophases are a way to show that the missions of the prototypes tend to evolve over time in a standard way. The early prototypes tend to focus on key concepts, such as requirements, architecture, or technology. The next several prototypes introduce and focus on the secondary concepts of requirements, architecture, and technology. After that, the focus shifts to design and implementation concerns. The last set of prototypes emphasizes optimization and deployment (in the target hardware and in the customer's environment). The shift in focus of the prototypes tends to be gradual, hence the overlapping nature of the macrophases. If required, Preliminary Design Reviews (PDRs) and Critical Design Reviews (CDRs) are easily incorporated into the process. Usually a PDR occurs at or near the end of the first macrophase and a CDR occurs at or near the end of the second macrophase.

Each macrocycle contains several microcycles, or spirals. Each microcycle is fairly short, usually completing within 4–6 weeks. Each microcycle is focused around the production and delivery of a single incremental prototype with limited but high-quality functionality. This is most commonly focused around one or a small number of use cases, but may also include specific risk-reduction activities.

Within the microcycle, the developers work to produce the high-quality object collaborations that realize the use cases of the prototype. During this process, the increasingly complete collaborations are executed dozens to hundreds of times. This very short execution cycle—in the order of minutes to hours (at the long end)—is called the nanocycle. If an object collaboration will ultimately consist of 100 objects, experience has shown—clearly—that the best way is NOT to put down all 100 objects and say "Oh God, I Hope This Works"[3] but instead to start with one (incomplete) object and get that to work in isolation, through model execution. Then add another object and get them to work together. Then refine the objects, or add a third; and so on, making small enhancements to the capabilities supported in the collaboration but validating, through execution, each small incremental step. Executable modeling tools, such as Rhapsody™, make this process highly efficient. The basic premise of the nanocycles is to make tiny incremental steps and demonstrate through execution that they are right before adding the next. The so-called "agile processes" such as the Extreme Programming (XP) approach focus almost exclusively on the nanocycle scale of development.

---

[3]   Prayer might be a wonderful thing, but is outside the scope of this book. My observation is that it does not lend itself to the development of high-quality software. ;-)

Harmony Process Overview

Note: The version of the Harmony process described in the remainer of this chapter is version 1.5; it is under configuration management at Telelogic so that users can be sure that they have a coherent set of artifacts that are internally consistent. It is anticipated that over time, the process will be further modified and updated.

The Harmony process is a general systems-development process that, while emphasizing the real-time and embedded-software development aspects, includes the steps to produce general-purpose software and systems. The Harmony process has been used effectively on very small 1–3 person projects as well as large teams consisting of hundreds of team members. Harmony is a highly scalable "medium-weight" process, striking a balance between static heavyweight processes and lightweight, so-called "agile methods" such as Extreme Programming (XP), while incorporating aspects of both.[4]

The Harmony process comes in two generic forms. The first is intended for projects that are larger in scale and have significant hardware-software codevelopment. Because of the long lead times necessary for the development of mechanical and electronic components, it is important that all the requirements be fully described

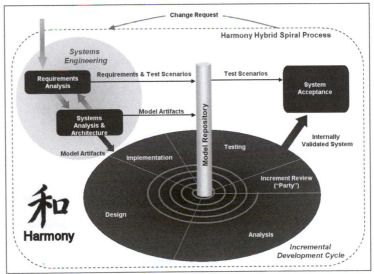

Figure 2.3 General Harmony Hybrid-spiral

---

4   The "nanocycles" timeframe in the process corresponds to the agile method's primary scale of concern.

and understood and the overall architecture be well defined before any significant design occurs. For this reason, the general Harmony Process Macrocycle is a hybrid of the classic "V" cycle and a spiral, as shown in Figure 2.3.

The general form of the Harmony process shown in Figure 2.3 has an upfront effort—referred to generically as "systems engineering" in which the requirements are fully specified and organized into use cases, the subsystem architecture is defined, the requirements are allocated to the subsystems, and, at the subsystem level, requirements are allocated to the engineering disciplines of mechanical, electronic, chemical, and software.

The systems engineering portion of the general Harmony Hybrid-spiral was developed primarily by Dr. Hans-Peter Hoffman, Chief Systems Methodologist for Telelogic. He and I worked together for a number of years to create a fully integrated systems and software process; the result of that work we called the Harmony process because it harmonizes the systems and software engineering disciplines together into a single coherent process. The three phases in the systems engineering part of the process are shown in Figure 2.4.

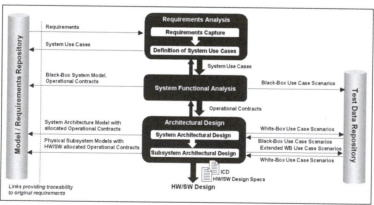

Figure 2.4 Systems engineering phases of the Harmony Hybrid-spiral

After the systems engineering phases, the incremental development cycle can begin. At this point, the two variants of the process, General Harmony and Harmony-SW, are remarkably similar. There is an important difference in the analysis phases: in the general process, previously detailed use cases are selected and used as the basis for the prototype development, while in the software-only process, the as-yet-unspecified use cases must be detailed as a part of the spiral. Other than that, the spirals are essentially identical.

The spiral part of the process is shown in Figure 2.5.

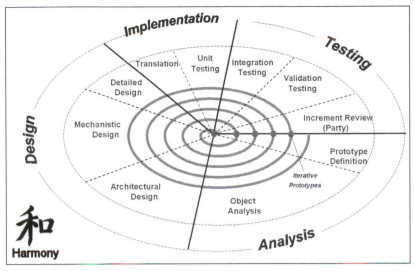

Figure 2.5 Harmony spiral (overview)

In the General Harmony process, the prototype definition phase within analysis is simply a matter of selection of the use cases previous specified in the systems engineering work. In the software-only spiral model, these use cases have been identified (named and given a single paragraph mission statement) but the detailed requirements for those use cases have not yet been specified. In this latter case, the first part of the spiral details the use cases so that they are fully specified.

The *next* two sections provide the workflows for each of these phases. UML activity diagrams are used to show process activities, flows, and artifacts.

The Systems Engineering Harmony Workflows in Detail

The Harmony Microcycle has been discussed from an overview perspective, but in order to understand how to use the process, it is necessary to understand in more detail the work activities and artifacts produced.

System Functional Analysis

Figure 2.6 shows the system functional analysis workflow. It basically specifies that the use cases are taken, whether one at a time or simultaneously, and a use case model is constructed for each. This means that an executable model is constructed using semantically complete modeling provided by the UML. The details of how that is done are discussed in the next section. Each use case is validated via execution to ensure that it is complete, correct, consistent, and accurate.

This can be done incrementally—that is, the use-case model becomes increasingly complete and use cases are added to it, and the entirety of the model is validated at each step—or it can be done as separate use cases, then added together later. If the latter approach is used and the use cases are not fully independent, then it is possible for inconsistencies among the use cases to arise. In that case, a use-case consistency analysis is done by adding the use cases together into a single requirements model and executing that model as an integrated unit.

The reader should note that we will use objects to represent the use cases, and detail their interactions with sequence diagrams and specify their behavior with state machines and/or activity diagrams. The fact that we are using these semantically precise languages for modeling does not mean that we are doing design! This is a common misunderstanding by many people. The use of a semantically precise language, such as state machines, has nothing to do with what we are saying, merely that we are using a precise language to say it. In this context, we use semantically precise languages to specify the requirements but say nothing (yet) about design or implementation concerns.

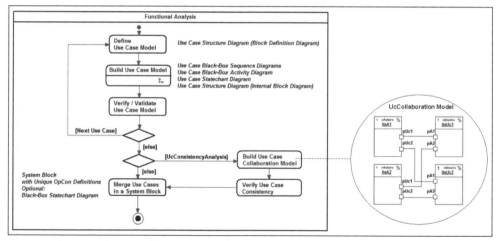

Figure 2.6 System functional analysis workflow

Build the Use-Case Model Workflow

One of the steps in the previous workflow (Figure 2.6) was "Build the Use-Case Model." That step is detailed in the next workflow (Figure 2.7). The workflow shows three alternative approaches that represent personal preferences. By the end of the workflow, you'll have created sequence diagrams showing the typical and exception interactions of your system with its environment, a summary of those sequences in an activity diagram, and a state machine providing an executable behavioral model of an object that represents the system use case.

Why represent the use case as an object? Use cases are themselves Classifiers in the UML and can have behavioral dynamics, such as state machines. However, you can show neither interfaces nor ports on use cases. Thus, we find it convenient for technical reasons only to model the use case with an object for this purpose. The object is nothing more than a notational convenience and can be thought of as a "use case" with a different notation.

This analysis is called "black box" because the internal structuring of the system isn't known or used at this time. In the next workflow, we'll go "open box," identifying subsystems and allocating functionality to them.

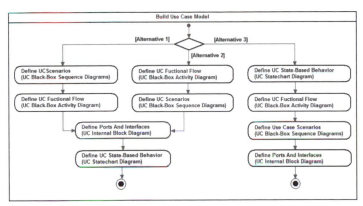

Figure 2.7 Build use-case model workflow

System Architectural Design

The next workflow is to specify the overall system architecture. This is done by identifying coherent functional blocks, represented as subsystem objects, along with their connection points (ports) and interfaces. The operational contracts ("op cons") are allocated to these subsystems. At this point, each subsystem is still "mixed discipline"—that is, it contains elements from various engineering domains, such as electronics, mechanical, and software. These subsystems are validated by executing them together and showing how they collectively reproduce the very same "black box" scenarios specified in the previous workflow. Note that "op cons" in the figure refers to operational contracts (service specifications in interfaces), BB is "black box" and WB is "white box" (i.e., subsystem level).

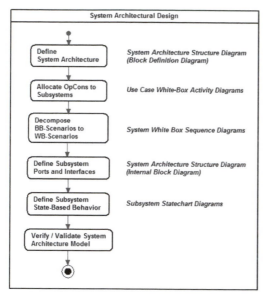

Figure 2.8 System architecture design workflow

Subsystem Architectural Design Workflow

The last workflow described in this brief process overview is subsystem architectural design (see Figure 2.9). The services are allocated to various discipline-specific components (mechanical, electronic, software, and so on); if they are not met by a single engineering discipline, then they must be decomposed (usually with an activity diagram) until they do. The interfaces between these components is specified at a high level but will be detailed more fully (e.g., port or memory addresses, bit-encoding, pre- and post-conditions) once the process enters the spiral portion of the process.

This workflow has three entry points. The first is to not create subsystem-level use cases but just to work forward from the allocated operational contracts (alternative 1 in the figure). The second entry point starts with the system-level use cases and decomposes them with «include» dependencies. Each use case is decomposed into a set of use cases, each of which will be satisfied in its entirety by one subsystem. Generally, each system use case decomposes into one or more use cases for each subsystem. This process is repeated for each system-level use case. At the end of that effort, each subsystem has a set of use cases that it must fulfill so that the system can fulfill its use cases. The last entry point (alternative 3) is a "bottom up" approach in which the set of operational contracts are clustered together into coherent units (use cases). I personally prefer alternative 2 but if the subsystems

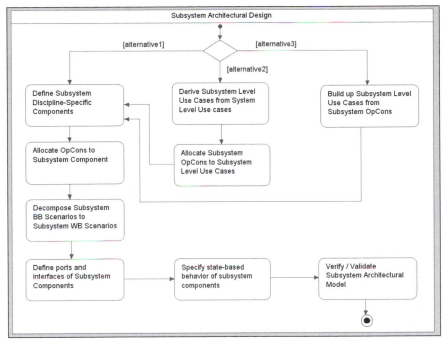

Figure 2.9 Subsystem architecture design

are simple, alternative 1 might be adequate. Other engineers may prefer alternative 3 when the subsystem is complex enough to warrant having its own use cases but prefer not to work top-down.

The Incremental (Spiral) Development Workflows in Detail

At the end of the workflows in the previous section, the model is handed off to the interdisciplinary subsystem teams and work enters the incremental development cycle (aka, microcycle or spiral). In this section, we detail only the software workflows of the spiral, but the reader should understand that, in the general Harmony hybrid-spiral, engineers of other disciplines are concurrently working in an incremental fashion as well. Integration occurs in the testing phase of the spiral, and this integration may include mechanical, chemical, electronic, and software disciplines. There may not be hardware components to integrate in any specific spiral, but there often is. It might be breadboard or wire-wrapped electronics, with mock-up or hand-built mechanicals, or first-run factory electronic and mechanical components. The point is to avoid the shotgun integration at the end of the project and plan for incremental integration as early as possible.

Increment Review (Party!) Workflow

The spiral starts in the increment review phase (also known as the party phase[5]). This phase is where the primary project planning and on-going assessment activities take place. Remember that there are two forms to the Harmony process. In the general form, the first time into the spiral, the software, the general schedule, software development plan, configuration management plan, reuse plan (if any) are defined. In the Harmony-SW variant, project scope and engineering approach are also selected and defined, and the system use cases are identified and given a one-paragraph mission statement. However, in this latter case, the use cases are not detailed—that takes place in the analysis phase of the spiral (described in the next section).

In subsequent spirals, the project and system are assessed against those plans and the plans modified as necessary. The primary artifacts assessed during the party phase are:

- Schedule

- Architecture

- Process

- Risks

- Next Prototype Mission

One of the more serious project management mistakes made is inadequate assessment and adjustment of projects during their execution. As DeMarco and Lister note, "You cannot control what you do not measure.[6]" It is equally important that you apply the measured information to make adjustments. In terms of schedule, such adjustments will be things like reassignment of resources, reordering activities, deletion of activities, reductions (or enhancements) of scope and/or quality, rescheduling subsequent activities, and so on.

Because the selection and implementation of a good architecture is crucial to the long-term success of a project and product, the party phase evaluates architecture on two primary criteria. First, is the architecture adequately meeting the needs of the qualities of services that are driving the architectural selection? Second, is that architecture scaling well as the system evolves and grows? The process of reorganizing

---

[5]   The party phase corresponds to both initial concept and post-mortem assessment phases in some other development process models. The use of the term "party" is to reinforce the notion of "celebration of on-going success" rather than a post-mortem "figure out why it died" analysis.

[6]   DeMarco and Lister, *Peopleware: Productive Projects and Teams*, New York, New York, Dorset House Publishing Company, 1987.

the architecture is called *refactoring the system*. If the project team finds that the architecture must be significantly refactored on each prototype, then this is an indication that the architecture is not scaling well, and some additional effort should be given to the definition of a more scalable architecture.

Early on in the project, selections are made about how to manage the project—what tools will be used, where they and their data are located and how they are accessed, security procedures, artifact review and quality assessment procedures, work and artifact guidelines, etc. The Party phase seeks to improve the efficiency of the process during the project by actively looking for and correcting problems and issues.

In my experience, the biggest single reason for project failure is ignoring risks. To manage risks, we recommend each project maintain a risk management plan. In this plan, each risk is identified and ranked and, where appropriate, a risk-mitigation strategy is described. Most of these will be activities to be done in the spirals to explore, reduce, or handle the risk. In the Party phase, the risk management plan is reviewed and newly identified risks are added.

Lastly, although the plan for the prototype mission is decided early on (and scheduled against), this plan is reviewed and possibly adjusted each iteration. It is common to make minor adjustments to the mission scope but, if nothing else, explicitly reviewing the plan ensures everyone knows what to do over the next 4–6 weeks it takes to complete the microcycle.

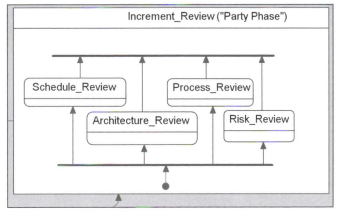

Figure 2.10 Increment review (party phase)

Analysis with the Harmony Process

The purpose of analysis is to define the essential properties of the system to be developed. The use of the term "essential" means that it defines the properties that, if missing, indicate the system is wrong or incomplete. In model-driven architecture (MDA)[7] terms, we are constructing a platform independent model (PIM) of the prototype capabilities in the analysis phase. The two primary workflows in the analysis phase of the spiral are the prototype definition and object analysis.

Prototype Definition Workflow

In the General Harmony form, this workflow is rather trivial—collecting the already completely defined use cases that will be added to the prototype in this iteration. In the Harmony-SW form, the use cases have been identified but not detailed, and so at this point the use cases being realized in this prototype must be completely detailed (the ones that will be realized in later prototypes are ignored for now). We will focus just on the Harmony-SW form in this section. The workflow for this effort is provided in Figure 2.11.

In this phase, the requirements of the current prototype are identified and captured in detail. The use cases for the prototype have already been identified, but the detailed specification of what the use cases contain has not yet been created.

There are two primary ways to detail a use case: by example and by specification. By "by example," we mean that a (possibly large) set of scenarios is created that illustrates typical and exceptional uses of the system for the use case in question. The advantages of scenarios are that they are easy for nontechnical stakeholders to understand and they can serve as a basis for the set of test vectors to be applied later to the completed prototype. The disadvantages of scenarios are that requirements of a use case are spread out over possibly dozens of different sequence diagrams rather than being in a single place, and the requirements may be difficult to represent concisely. Additionally, some requirements, such as "The tank shall be painted with a green camouflage-scheme," are not really behavioral. They are merely characteristics that are either true or not, of the resulting system.[8] Scenarios are almost always represented with UML sequence diagrams.

The other approach to detailing use cases is "by specification." This specification may be informal, using text to describe the requirements of the use case, or a formal

---

[7]   See *www.omg.org* for a set of standards that define MDA.

[8]   Such requirements are called "system parametric" requirements and trace not to use cases (as do functional and quality of service requirements) but instead to the System object.

behavioral language such as UML state machines or UML activity diagrams. The advantages of detailing use cases by specification are that it is concise, it can be made more precise than scenarios typically are, and it is easy to represent requirements that are difficult to show in scenarios. The disadvantages are that it is more difficult, particularly for nontechnical personnel, to understand, and directly relating the requirements to the design may also be more difficult. For continuous and piecewise continuous behavior required, we recommend using control law diagrams or activity diagrams to represent the continuous behavior of these individual use cases.

We find it best to use both informal text and formal languages together for the use-case specifications. Natural language is excellent at explaining "why" because it is both rich and expressive. However, it is also vague, ambiguous, and imprecise. Formal languages excel in precise statements about "what" is needed. A combination of a precise formal description, such as with a state machine, coupled with explanatory text is the best of both worlds.

Both exemplar and specification approaches are useful, and in fact the Harmony process recommends that both be used together. A formal specification using state machines or activity diagrams captures the requirements concisely, while scenarios derived from the formal specification can aid the nontechnical stakeholders in understanding the system. Further, the scenarios derived from the formal specification may be used to generate the test vectors for validation at the end of the microcycle.

Requirements[9] are detailed using a combination of:

- Sequence diagrams
- State machines
- Activity diagrams
- Control law diagrams (non-UML)
- Textual descriptions
- Quality of service (QoS) constraints
- (SysML) requirements diagrams

---

[9] The UML does not have the notion of a "requirement" as a first-order concept. In the Systems Modeling Language (SysML), a recently released UML profile specialized for systems engineering, requirements elements are explicitly representable. For more detail, see *www.omg.org*, *www.ilogix.com*, or *www.telelogic.com*.

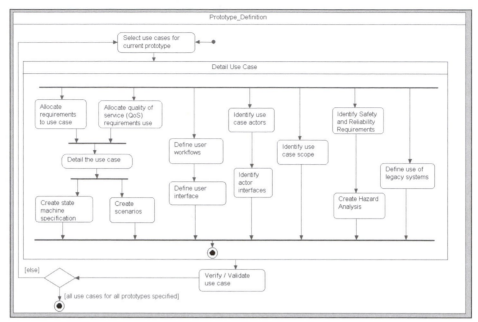

Figure 2.11 Prototype definition in the Harmony-SW process variant

Object Analysis Phase

A use case can be thought of as a bag that contains a set of detailed requirements relating to a single system capability or operational usage. The realization (implementation in UML-speak) of a use case is a collaboration, a set of objects working together to achieve this coherent set of requirements.

Object analysis in the Harmony process constructs this collaboration of essential objects, and is performed a use case at a time. This means that for the current prototype, one collaboration is constructed for each use case implemented by the prototype. In MDA terms, the essential model is called the platform independent model (PIM). The Harmony process constructs the PIM in an incremental fashion, one (or a few) use case(s) at time. This is illustrated in Figure 2.12.

In the Harmony process, the "Here be dragons" step is labeled "Apply Object Identification Strategies." Table 2.1 lists and briefly describes these strategies. We have found these strategies to be a remarkably effective way to identify the essential classes and objects within a collaboration. The application of these strategies will be the subject of Chapter 5.

Table 2.1 Object discovery strategies

Strategy	Description
Underline the noun	Used to gain a first-cut object list, the analyst underlines each noun or noun-phrase in the problem statement and evaluates it as a potential object.
Identify causal agents	Identify the sources of actions, events, and messages; includes the coordinators of actions.
Identify services (passive contributors)	Identify the targets of actions, events, and messages as well as entities that passively provide services when requested.
Identify messages and information flow	Messages must have an object that sends them and an object that receives them as well as, possibly, other objects that process the information contained in the messages.
Identify real-world items	Real-world items are entities that exist in the real world, but are not necessarily electronic devices. Examples include objects such as respiratory gases, air pressures, forces, anatomical organs, chemicals, vats, etc.
Identify physical devices	Physical devices include the sensors and actuators provided by the system as well as the electronic devices they monitor or control. In the internal architecture, they are processors or ancillary electronic "widgets." This is a special kind of the previous strategy.
Identify key concepts	Key concepts may be modeled as objects. Bank accounts exist only conceptually, but are important objects in a banking domain. Frequency bins for an on-line auto-correlator may also be objects.
Identify transactions	Transactions are finite instances of associations between objects that persist for some significant period of time. Examples include bus messages and queued data.
Identify persistent information	Information that must persist for significant periods of time may be objects or attributes. This persistence may extend beyond the power cycling of the device.
Identify visual elements	User-interface elements that display data are objects within the user-interface domain such as windows, buttons, scroll bars, menus, histograms, waveforms, icons, bitmaps, and fonts.

Table 2.1 Object discovery strategies (continued)

Strategy	Description
Identify control elements	Control elements are objects that provide the interface for the user (or some external device) to control system behavior.
Apply scenarios	Walk through scenarios using the identified objects. Missing objects will become apparent when required actions cannot be achieved with existing objects.

Care should be taken to minimize the introduction of design elements during analysis. Limit the collaboration at this point to elements which clearly must be present in the object analysis model. For example, if the collaboration is to model the use case "Manage Account" for a banking system, then if the collaboration does not contain objects such as Customer, Account, Debit Transaction and Credit Transaction, then you'd say it was wrong. In a navigation system, you would expect to see concepts, represented by objects or their attributes, such as Position, Direction, Thrust, Velocity, Attitude, Waypoint and Trajectory. The goal is to include only the objects, classes, and relations that are essential for correctness and not to include design optimizations.

A key question arises during the construction of the object collaboration: "Is this right?" Are the concepts properly represented? Are the relationships among those concepts correct? Do they behave appropriately? The answer to these questions is answered rapidly during the nanocycle. You can see the Harmony Spiral Nanocycle activity in Figure 2.12. The idea is to make tiny incremental changes and then quickly execute the collaboration to make sure that you got it right. You can really only evaluate the correctness of an object model via execution and test. With executable modeling tools such as Rhapsody, this is very fast and easy.

The nanocycles consist of generating and executing the object analysis model while it is in various stages of completion, rather than waiting until the end. Testing becomes a continuous process rather than something done only at the end, resulting in higher-quality systems with less effort and in less time. Take the sequence diagrams used to show requirements scenarios, elaborate them with the objects just created and demonstrate, via execution, that they fulfill the expected roles within that scenario realization. This is the key concept behind agile methods, such as extreme programming—make tiny steps and validate them before you move on.

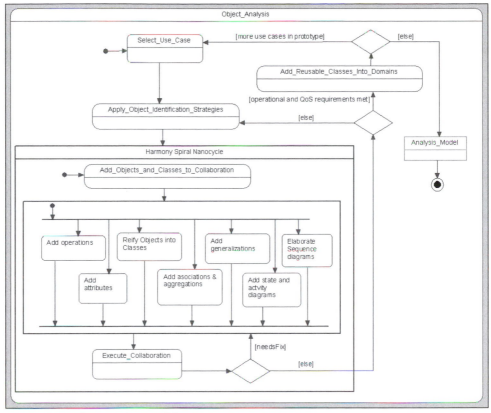

Figure 2.12 Object analysis workflow

Design with the Harmony Process

An analysis model defines a coherent set of required properties of the system under development. These are represented as a set of use cases and its details (e.g., sequence diagram, state machine, activity diagrams, quality of service constraints) and as a collaboration of essential objects roles whose correctness is verified through execution and test. The object analysis model is driven primarily by the functional requirements and is demonstrated to be functionally correct. It is not, in general, optimal. That optimality is introduced in design.

A design model is a concrete blueprint for exactly how the essential properties will be realized. An analysis model may be implemented by many different designs with different optimization characteristics. While an analysis model presents a set of possible solutions, a design is a particular solution to the problem. Design is always an optimization of an analysis model. The design process is basically:

- Identify the design (optimization) criteria

- Rank the design criteria in order of criticality
- Identify design patterns or other solutions that optimize the more important of the design criteria at the expense of those of lesser importance.

Many of the design criteria are the quality of service constraints from the analysis. There may be others as well, such as reliability and safety level, reusability, maintainability, simplicity, time to market, and so on.

In the Harmony process, we do this at three levels of abstraction. The architectural level of abstraction optimizes the system at an overall gross level. As we will see, there are five (sometimes six) views of the architecture that are optimized more-or-less independently. The mechanistic level of abstraction focuses on collaboration-wide optimizations, where a collaboration realizes a single use case. The scope of concern here is an order of magnitude smaller than architectural scope. Lastly, the detailed design is at the level of the individual class or object. At this level, individual objects are optimized and focus is centered on the 5% (or so) of the objects that have special optimization concerns.

Architectural Design Phase

As mentioned earlier, the Harmony process recognizes five (or six) important views of architecture:

- Subsystem and Component View
- Concurrency and Resource View
- Distribution View
- Safety and Reliability View
- Deployment View
- (optional) Security View

In the Architecture Design phase of the spiral, one or more of these views is elaborated, depending on the needs of the current prototype. This is done primarily via the application of architectural design patterns (see the author's Real-Time Design Patterns[10] for an in-depth presentation of patterns in each of these architectural views). These patterns are large in scope, affecting most or all of the system.

Architectural design representation uses the same UML diagrams as in systems architecture and object analysis—class diagrams to represent the structure and

---

[10] Douglass, Bruce Powel, *Real-Time Design Patterns: Robust Scalable Architecture for Real-Time Systems*, Addison-Wesley, 2002.

sequence diagrams to represent collaborative behavior, state machines to model the behavior of individual elements, and sequence diagrams to represent the collaboration of groups of such elements.

The subsystem and component view identifies the large scale architectural pieces of the system, their responsibilities, and their interfaces among each other and to the external actors. The concurrency and resource view identifies the concurrency units (modeled as «active» objects), the policies for scheduling those concurrent units and how they synchronize and share resources. The distribution view identifies how objects are distributed across multiple address spaces, how they find each other, and the means and protocols they use to interact and collaborate. The deployment view shows how the products from the different engineering disciplines work together. While the UML provides deployment diagrams, the SysML prefers to use class diagrams to represent the deployment architecture because they are richer and more expressive than deployment diagrams. Lastly, the security architecture is important in some systems and has to do with the policies and procedures to maintain data integrity and security.

Figure 2.13 shows the workflow for architectural design. You can see in the figure how the design criteria are identified and ranked and then used to select appropriate

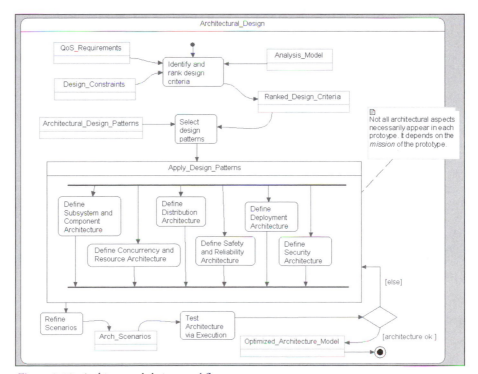

Figure 2.13 Architectural design workflow

architectural design patterns. The parallel nature of the applications of the patterns in the different aspects of architecture emphasizes that 1) the order in which they are introduced is a personal preference and 2) not all the aspects of architecture must be realized in the same prototype. It is common to put off some aspects of the architecture until later prototypes when they are low risk and concentrate on the high-risk aspects in earlier prototypes.

Mechanistic Design Phase

The mechanistic design phase is concerned with the optimization of individual collaborations, each realizing a single use case. The scope of mechanistic design decisions is generally an order of magnitude smaller than those found in architectural design, since a system typically consists of one to several dozen use cases. Similar to architectural design, mechanistic design largely proceeds via the application of design patterns, although the scope of the patterns is much smaller than that found in architectural design. This is where the classic "Gang of Four" patterns[11] and other, more fine-grained patterns are applied.

The mechanistic design view is an elaboration of the object analysis view and uses the same graphical representation—class and sequence diagrams for collaborations structure and sequence, activity, and statechart diagrams for behavior.

The workflow for mechanistic design is shown in Figure 2.14. The flow is similar to architectural design in that the first step is to identify what you want to optimize (the design criteria), how important each of the criteria is, and then to find design patterns that optimize the more important of these at the expense of the least. The resulting design collaboration is tested, not only to make sure that you haven't broken the functionality in the original analysis collaboration, but also to ensure that you have achieved the desired optimization.

Detailed Design Phase

The detailed design phase elaborates the internals of objects and classes, and has a highly limited scope—the individual object or class. Most of the optimization in detailed design focuses on the issues of:

- Data structuring (space or time optimization)
- Algorithmic decomposition

---

[11]   Gamma, et. al, *Design Patterns: Elements of Reusable Object Oriented Architecture*, Addison-Wesley, 1995.

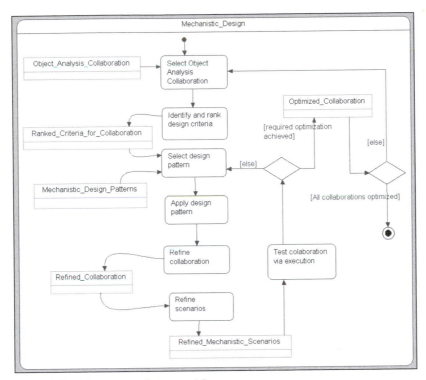

Figure 2.14 Mechanistic design workflow

- Optimization of an object's state machine
- Object implementation strategies
- Association implementation
- Visibility and encapsulation concerns
- Ensuring compliance at run-time with preconditional and postconditional invariants (such as ranges on method parameters)

There are many rules of thumb, guidelines and practices for detailed design, although these commonly fall under the title of "idioms" rather than "patterns." For most objects, detailed design is little more than a trivial detail, but there is usually a small (5% is typical), but important, set of objects that require special attention during detailed design. Figure 2.15 shows the detailed design workflow.

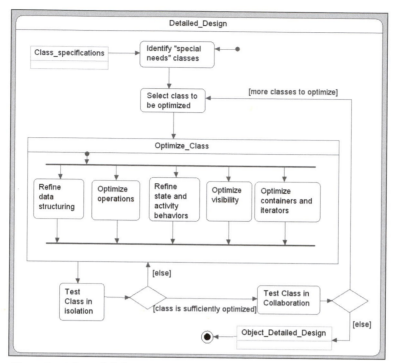

Figure 2.15 Detailed design workflow

Implementation

The implementation phase is concerned with the correct construction of the properly working architectural elements. This phase includes the generation of code (whether it is automatically generated from your model, written by hand, or a combination of the two), unit level testing of that source code and the associated model elements, integration with legacy source code, the linking together of the pieces of the architectural element (including, possibly, legacy components), and model-based peer review of the architectural element itself. It should be noted (since the question comes up all the time), that it is not a problem to integrate existing legacy code during the implementation phase. It is, in fact, unusual when this is not the case.

The primary artifacts for the implementation phase are:

- Source code generated from the model elements
- Unit test plan, procedures and results (textual documents)
- Inspection report for the source code (textual document)
- Compiled and tested software components

The straightforward workflow is shown in Figure 2.16.

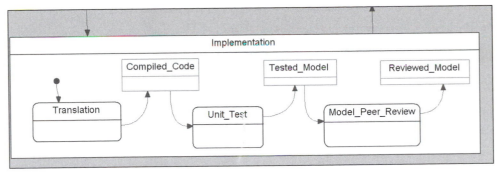

Figure 2.16 Implementation workflow

Test

The test phase constructs the prototype from the architectural elements and ensures that they fit together (integration), and that the prototype as a black box meets its mission statement (validation). The first of these, integration, is concerned with the construction of an integrated architecture from the architectural pieces constructed in the previous phase. The tests are limited to demonstrating that the interfaces of the architectural elements are used properly and none of the constraints are violated. This normally proceeds in a stepwise fashion, according to an integration plan, that adds the architectural elements one at a time. It is in this phase that hardware elements are formally integrated with the software elements for prototypes that have hardware-software integration as part of their mission. The integration test plan and procedures may be developed once the subsystem and component architecture of the prototype is specified—that is, either at the end of systems engineering or architectural design. Figure 2.17 shows the integration workflow that is done at the end of each spiral to bring together the architectural elements produced in that build of the system.

The validation phase tests the assembled prototype against its mission. The mission for a prototype is normally a small set of use cases and/or the reduction of a small number of risks. The validation test plan and procedures may be written as soon as the requirements for the prototype are understood—that is, at the end of the microcycle's prototype definition workflow.

If defects are found during testing, they may be either fixed then (required if the defect is severe enough) or may be deferred until the next prototype.

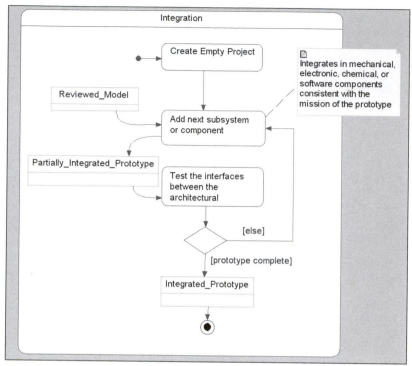

Figure 2.17 Integration workflow

The primary artifacts for the test phase are:

- Integration test plan, procedures, and results
- Validation test plan, procedures, and results
- Tested, executable prototype
- Defect report

It should be noted that the Functional and QoS test vectors are primarily composed of the use-case sequence diagrams specified in the Analysis phase (Harmony-SW) or the System engineering phase (General Harmony). The regression test vectors are a subset of the test vectors from previous microcycles. Additional tests, such as stress, volume, coverage, and fault-seeding tests, are added manually. The validation workflow is shown in Figure 2.18.

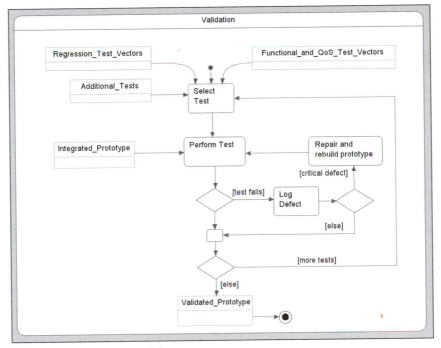

Figure 2.18 Validation workflow

Summary

This chapter briefly described the Harmony process, in both the general form (that contains a systems engineering phase) and software-only form. The Harmony process consists of a set of integrated workflows for each phase in the process. In the general form, the development spiral is preceded by a systems-engineering phase that fully elaborates all the requirements of the system and constructs a well-formed and executable use-case model. Then an internal subsystem architecture is created and the requirements, represented primarily as "operational contracts," are allocated to these subsystems, usually being grouped together into subsystem-level use cases. The subsystems themselves are decomposed into high-level engineering discipline-specific components for electronic, mechanical, chemical, and software aspects and the logical interfaces among these components are defined. At this point, the models are handed off to the subsystem teams and their engineering disciplines for further development.

In the Harmony-SW variant, some optimizations are enabled by the lack of significant hw–sw codevelopment. Rather than fully specify all requirements up front in Harmony-SW, use cases are identified but not detailed until they are about to be developed. With the development spiral, a small set of use cases are selected and, at that time, are detailed with sequence diagrams, state machines, activity diagrams, etc. Then an object analysis model is created that correctly captures the functional requirements of the use case and this model is validated via model execution.

This analysis model requires optimization and the introduction of specific realization technologies, which is done in design. Design is driven by the optimization criteria such as the quality of service requirements. Design occurs at three levels of abstraction: architectural design optimizes the system at an overall level, mechanistic design optimizes collaborations of objects, and detailed design optimizes individual objects.

Implementation produces high-quality architectural units (e.g., subsystems or components) that are unit-tested and peer reviewed. These components incorporate legacy code and components as appropriate. Integration brings together these high-quality components and integrates them—along with relevant hardware elements—into an integrated prototype. Validation demonstrates the correctness of the prototype against its mission (requirements and risks to be reduced). Identified defects are then either fixed in the testing phase of the current spiral (for critical defects) or at some appropriate point in the subsequent spiral.

The next spiral continues where this one left off, adding in (and detailing, in the Harmony-SW variant) the next few use cases and risk mitigation items. The incremental prototype becomes increasingly more capable and complete as time goes on, until finally all the requirements have been realized and validated. At this point, the prototype is released to manufacturing or the customer.

In the next several chapters, we will introduce problems and exercises for you to gain experience in applying these workflows to real-life applications. Two applications are presented. Appendix A provides a problem statement for a traffic-light control system. This is a small-scale application and demonstrates how the process workflows apply to systems of moderate size. Appendix B presents a much larger-scale system—an unmanned air vehicle—complete with ground and airborne aspects. This larger system allows the exploration of problems of scale and architecture that don't arise in small systems.

In both cases, the purpose of the problems is to give you a context in which to gain experience by doing. It is not to construct a fully complete and validated model of those systems. This is especially true of the unmanned air vehicle system. To create and explain a full UAV system would take thousands of pages. However, the scope of the problems provides rich fodder for using model-based approaches to development.

Specifying Requirements

What you will learn:

- How to identify types of requirements
- How to group requirements together
- How to identify use cases
- How to capture and represent quality of service requirements
- How to identify operational scenarios
- How to use state machines for representing requirements
- How to capture complex requirements
- How to capture algorithmic requirements with activity diagrams
- How to validate requirements models

Overview

One of the most important things that engineers must do is specify requirements on systems. This is a crucial step in any high-reliability process but, unfortunately, is often poorly executed. Few engineers have been trained in the capture and management of requirements, and this lack results in huge "Victorian novel" style requirements documents being constructed, which are incomplete, inconsistent and ambiguous. Because such documents are captured in text, they inherently lack the precision necessary for validating them prior to initiating design. This means that the quality of the requirements cannot be assessed in any reliable way. Requirements defects are one of the most important kinds of strategic flaws in systems and are certainly the most expensive to repair because they are introduced early, (typically) identified late, and have far-reaching implication and scope. The UML and the Harmony process offer a better way.

The UML only implicitly represents requirements, but does explicitly represent coherent requirements sets via the concept of use cases. A use case can be thought of as an operational view of the system clustered around some generic kind of system usage. As such, it implicitly "contains" a coherent set of requirements clustered around that system usage. More specific uses of the system—called *scenarios*—are highly specific "paths" through the use case. From the DODAF and MODAF point of view, use cases and scenarios can be part of the "operational architecture" of a set of interacting operational nodes. In operational perspective, the use cases and scenario views are vital to understanding how the operational nodes (that will be met by actual systems[1]) collaborate with other elements in the context of the system environment.

We can also model the requirements from the specification, rather than the operational, perspective. These are statements of static and dynamic properties of the system—the "system architecture" in DODAF/MODAF terms. When we look at requirements from this perspective, precise, unambiguous statements of system properties and aspects must be made. Formal languages within the UML, such as statecharts and activity diagrams, are tailor-made for such precise specifications.

The question arises as to how to best represent requirements from the operational and specification perspective. If you didn't know how to do this in a language as commonplace as English, how much more daunting it must be to do it with a less common language, such as the UML!

This chapter provides a set of exercises that will give you practice and experience representing requirements, from both the operational and specification point of view. The answers to the exercises are given in Chapter 8. Be aware that there is no one "right" answer—there are many good ways to model requirements so that they are clear, complete, consistent, and unambiguous. There are many more bad ways to model requirements but, nevertheless, there are many good ways, so if your answers differ from those in the answers section of the book, it doesn't mean that your answers are necessarily wrong. In my experience, however, use cases are the most misused element in the entire UML. I believe this is not because use cases are inherently difficult to understand, but because few engineers are trained in capturing requirements. Most engineers end up trying to do what they are trained in—that is, design—with use cases, rather than capture requirements in a design-free way.

---

[1]   An operational node can be thought of as a role in the operational architecture, such as "Recon Platform," which, when the mission is actually run, is realized by a physical system such as "Satellite" or "AWACS."

Note: *Remember—all the problems in this book are based on two example problem statements; both are given in appendices of this book. Appendix A details the requirements for the Roadrunner Traffic Light Control System while Appendix B details the requirements for the Coyote Unmanned Air Vehicle (CUAV). The solutions for these problems can be found in Chapter 8.*

Problem 3.1 Identifying Kinds of Requirements for Roadrunner Traffic Light Control System

The UML does not have an explicit concept called a "requirement." Even though use cases, sequence diagrams, state machines, activity diagrams, and constraints of various kinds are used to model the properties of requirements, the requirement concept itself is lacking. For this reason, the Systems Modeling Language has added one, and Rhapsody lets us add requirements into the models.

A useful way to think about requirements in the UML is that they are model elements that specify some aspect of a system from an external, rather than internal, view of that system. Note that requirements are not classes, even though they both exist only at design time, because classes instantiate to objects, but requirements do not. A requirement is a kind of "correctness" constraint that applies to a system.

I believe that there are many different kinds of requirements, as shown in Figure 3.1. An operational requirement is one that specifies how the system is to collaborate with other elements (actors) in its environment. For example, if an unmanned air vehicle (UAV) must interact with a set of GPS satellites to navigate its terrain, how that interaction occurs is an operational aspect of that system. Functional requirements have to do with the behavior of the system. If a UAV must be able to point a gimbaled optical sensor at a point on the ground, then it is a function the UAV must be able to do. If functional requirement is a verb, then a quality of service requirement is an adverb since it specifies "how much." A quality of service requirement might be that our gimbaled optical sensor will be able to adjust in 0.1-degree increments with an accuracy of 0.01 degrees. A parametric requirement is both nonoperational and nonfunctional. Suppose that the UAV must be constructed so that it weighs less that 2100 pounds. The UAV's weight is a property of the system, which might be considered either static or dynamic, as the situation warrants. But it is clearly neither a behavioral nor an operational aspect, yet it is important to represent. Finally, a design requirement has to do with the characteristic of the design per se but not of the system as it exists in the field. An example design requirement might be that the UAV is designed so that the next UAV can reuse at least 70% of the model, or that the recurring cost of the system is less than $1.2M (not unreasonable at all for our Coyote UAV).

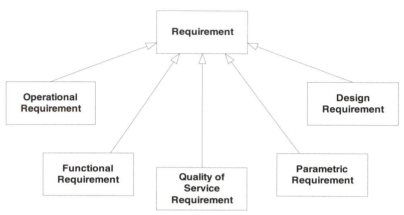

Figure 3.1 Requirement types

We model these requirements in somewhat different ways. For example, modeling reliable communications is very straightforward with a statechart, but how would you use a statechart to represent the fact that the nose of the UAV should be dark grey in color?

For this exercise, create a new model and add two diagrams (either a use-case diagram or object-model diagram will do). Label one "Overview Requirements" and the other "Special Mode Requirements." Take the text in the Roadrunner Traffic Light Controller problem statement in Appendix A, up to the "Configuration Parameters" section and put each requirement statement in a separate requirements element. Link the requirements elements together with dependencies as appropriate, using the following stereotype dependencies:

- «include» When a "whole" requirement includes a smaller "part" requirement

- «extends» When a smaller requirement specializes or adds to a main requirement

- «derive» When a requirement is a more detailed restatement of another

- «explain» When a comment explains a requirement.

Problem 3.2   Identifying Use Cases for the Roadrunner Traffic Light Control System

Use cases are functional uses of a system from an external operational standpoint. Remember, to be a use case, the following conditions must be true:

- It returns a value to at least one actor
- It contains at least three scenarios, each scenario consisting of multiple actor-system messages
- It logically contains many operational, function, or quality of service requirements
- It does not reveal or imply anything about internal structure of the system
- It is independent of other use cases and may (or may not) be concurrent with them
- There should be no fewer than three and no more than three dozen use cases at the "high level"
  - "Large" use cases may be decomposed with «includes» and «extends» relations into smaller use cases
  - Use cases may be specialized with the generalization ("is-a-specialized-kind-of") relation

Use cases may be identified in many different ways. Some analysts prefer to identify operational scenarios and then group them together into sets related by common system usage. Some analysts prefer to identify separate system uses and then detail them with many scenarios. Still others prefer to identify the actors that interact with the system and concentrate on how each actor collaborates with the system.

Select one of these strategies and identify the use cases for the Roadrunner Traffic Light control system. Also identify the actors in the system context and draw the use-case diagram for the system. Once completed, apply the criteria provided above to make sure you've made a good selection of use cases.

Problem 3.3   Mapping Requirements to Use Cases

The UML itself does not provide the notion of a requirement per se, so we are free to add dependencies that we might need to relate these model elements. In addition to the stereotypes of dependencies between requirements, I also use the «specifies» dependency to show that a use case is partially specified by a requirement element. I draw this dependency from the use case to the requirement element.

Take one of the use cases identified in the previous problem and create a new use-case diagram for it. On this diagram use the dependency relations to relate the operational requirements related to that use case as a way of defining that use case. You may also add additional relations among the requirements themselves, as appropriate.

Problem 3.4   Identifying Use Cases for the Coyote UAV System

I've seen customer models that have a flat organization of 500 or more use cases. These models are practically impossible to use. Remember one of the previous rules of good use cases: *There should be no fewer than three and no more than three dozen use cases at the "high level."* When the system is highly complex, it becomes necessary to create taxonomies of use cases at different levels of abstraction. The bottom, or leaf, level use cases are still entirely necessary but the taxonomy allows us to view the requirements organization hierarchically as well as horizontally. In practice, I have found this a crucial aspect of useful use-case models for complex systems.

For this exercise, identify the use cases for the Coyote UAV system (see Appendix B). Because of the complexity of the system, it will be necessary to identify "high-level" use cases that are decomposed into more detailed use cases. Use the «include» stereotype of dependency to create a whole-part taxonomy of use cases at different levels of abstraction. Add a package to hold nonoperational requirements such as system weight and payload capacity. Stereotype this package with the «parametrics» stereotype on the package as a way to indicate that this will be a placeholder for system parametric requirements. Validate the use-case model against the guidelines given in Problem 3.2.

Problem 3.5   Identifying Parametric Requirements

For the Coyote UAV, identify the system parametric requirements. Remember that these are requirements that specify nonoperational aspects of the system, such as physical characteristics. Add these elements to the package created for this purpose in the previous exercise. Create a use-case diagram that shows this package containing the requirements elements.

Problem 3.6 Capturing Quality of Service Requirements in Use Cases

Use cases are coherent sets of functional requirements. Good use cases are named with strong verbs.[2] If use cases, and their contained requirements, are verbs, then quality of service (QoS) requirements are adverbs; that is, they modify the use case or the specific requirements within those use cases. They specify "how much," "how fast," and "to what degree." Such requirements are historically known as nonfunctional requirements, but these days they are more commonly known as QoS requirements. How are QoS requirements represented?

QoS requirements are themselves requirements and can be related to use cases or requirements. I use the «qualifies» stereotype on dependency to represent the relation between an element and a QoS requirement. This problem is for you to take a "Fly UAV" use case and, in a separate use-case diagram, add the QoS requirements that modify this use case and attach them with dependencies.[3]

Note that while Rhapsody provides its own requirements element (from SysML), it is quite common to use constraints to represent requirements elements. Such constraints (or requirements elements) can be shown on use-case diagrams or in the more detailed view of sequence, timing, and state diagrams, topics to be discussed later in this chapter.

Problem 3.7 Operational View: Identifying Traffic Light Scenarios

So far, we've been focused on requirements specification and identifying the use cases to which they belong. This is little more than a modernization of the traditional requirements approaches. This can be useful, but large numbers of requirements are still problematic—they are difficult to understand, ambiguous, and almost impossible to validate. Is it possible to use the modeling power of the UML to help us?

---

[2] As a naming convention, use cases should be named with strong verbs, although sometimes for highly reactive (stateful) systems, you'll see use cases named with the word "... mode" in the title.

[3] A common question I hear is whether a separate package should be created for all QoS requirements. In my mind the answer is "Clearly not." If a QoS requirement applies to a use case or a requirement within a use case, then that adverb should be positioned next to the verb it is modifying—that is, it should be placed into the same package as the use case. If it is a parametric requirement, then it should be placed in the package with the parametric requirement it qualifies.

The answer is a resounding YES (although I'll bet you could guess I'd say that ;-)). We call this "detailing the use case." First, we'll discuss the notion of scenarios as a means of looking at the operational requirements in a use case. In later problems we'll examine the use of formal languages (state machines and activity diagrams) for specification.

A scenario can be thought of as a specific sequence of inputs and outputs that represents a single path through a use case. A use case normally has many different scenarios—several dozen is not uncommon—that show combinations of different inputs and different sequences. Scenarios are useful because they relate the system capabilities with the expectations of how the system will interact with other elements in its environment (i.e., actors). Further, while domain experts will understand the flow of scenarios and be able to discern reasonable from unreasonable scenarios, they are normally at a loss when confronted with formal languages. Lastly, scenarios are valuable at every level of abstraction. At the System level, there is a single lifeline for the system (or the system use case)—all other lifelines are actors. At the subsystem level, the system lifeline is "opened up" and the internal parts (subsystems) can be used as lifelines as well. At the collaboration level, the subsystem lifelines are decomposed into their primitive (nondecomposable) object roles that collaborate to fulfill the detailed functionality required.

The most common representation for scenarios, by far, is sequence diagrams. However, timing diagrams and communication diagrams (formerly known as collaboration diagrams) are sometimes used, although usually for design-level scenarios only.

A common problem with the application of scenarios is that there are so many possible scenarios that people get bogged down, particularly with the so-called "rainy day" scenarios. These are scenarios where faults occur and the system must handle them in some fashion. My recommendation is to define the sunny-day "everything goes right" scenarios first, and then create the rainy-day scenarios.

How many scenarios? A minimal spanning scenario set for a use case has every operational and operational QoS requirements contained within that use case represented at least once. For faults, it is usually necessary to classify them into equivalence classes. An equivalence class represents a set of faults that are identified and handled using the same means. In an anesthesia machine, a fault created when the endotracheal tube dislodges from the patient's throat is in the same equivalence class as if the oxygen supply line disengages from the wall supply or if the oxygen supply fails. In all these cases, the fault is detected in the same way (e.g., lack of end-tidal pressure wave) and the action is the same (same alarm to the attending physician). Thus, it is only necessary to have a single scenario for "failed to inflate patient's lungs." I recommend that all of the specific faults that can give rise to the generic fault be listed in a constraint, but I don't believe it necessary or useful to show the same scenario many times, once with each possible specific fault.

For this problem, take a use case identified for the traffic-light system and create three different scenarios. Take care when identifying the actors—they should be objects outside the scope of the system that have significant interactions with the system during the execution of the use case. Also, the preconditions for each scenario must be explicitly specified. I like to put these in a note or constraint at the top of the scenario.

Problem 3.8   Operational View: CUAVS Optical Surveillance Scenarios

The Coyote UAV system is much larger in size and scope than the Roadrunner TLCS. Consequently, there are more use cases as well as a deeper layering of them. However, the same technique of expressing the operational aspects of the system can be used. It is, however, even more important to organize and target the scenarios better because there will be so many use cases, each of which will have many scenarios. This problem is to take the use case Perform Optical Surveillance and create three scenarios: in the first scenario, the payload operator moves the gimbaled video camera around with a joystick to examine the area over which the Coyote is flying. In the second scenario, the payload operator should locate a target visually, select the target in the visual field, and then command the camera to follow that target autonomously as the UAV flies in the area. In the third scenario, the payload operator manually scans the area and, upon finding a potential target, zooms in 3x on the target and then manually scans the surrounding area.

Problem 3.9   Specification View: Use-Case Description

Text is a very useful medium for capturing requirements—it is both expressive and readable. However, it lacks precision and is ambiguous and so is problematic if it is the only representation for requirements. In previous examples, we've used requirements elements to organize requirements into taxonomies. It is useful, however, to also provide a "structured text" description of the use case itself. This is normally done in the use-case description field in tools such as Rhapsody. The format I like to use for this is as follows:

Name: name of the user case

Owner: person or team in charge of developing the use case

Purpose: the user-purpose of the use case (what value it brings, what purpose it serves)

Requirements (optional): list of requirements (optional if requirements diagrams are used)

Data (optional): Data values inherently consumed, manipulated, or generated by this use case—need not be shown especially if information flows are specified

Preconditions: what must be true before the use case runs

Postconditions: what the system guarantees to be true when the use case completes

Reference documents: hyperlinks to relevant standards or specifications

Different authors suggest different formats to this standard use-case header. Some, for example, will include lists of actors or constraints. My recommendation is not to replicate information already present (or to-be-captured) graphically, so I don't describe the actors nor list the primary scenarios here.

This problem is to use this (or whatever standard format you prefer) to describe one use case for the Roadrunner traffic light control system ("Detect Vehicle") and one use case for the Coyote UAV.

Specification View: State Machines for Requirements Capture

Many systems exhibit "reactive behavior"; that is, the system waits for events of interest and them reacts to them by executing some set of actions, changing state, and then waiting for the next event. This kind of behavior is precisely what state machines excel at specifying. We will use such state machines to specify the requirements of reactive systems. We may (and most likely will) use state machines later in the design and implementation of the very same systems, but in this section we are entirely concerned with the appropriate specification of requirements and not the design or implementation.

Some engineers are surprised at this use of state machines in requirements specifications, but state machines have been used to help specify requirements of complex reactive systems for more than 30 years. Statemate™ from Telelogic (formerly I-Logix) is one example tool that was originally developed for military aerospace system requirements specification in the mid-1980s. It is still in common use in such environments. In Statemate, functional decomposition is applied to the system, and these functions are defined using a combination of statecharts, truth tables, and action statements. Statemate is used primarily by systems engineers for the specification of requirements, not by software or hardware engineers designing those same systems. UML can also model such specifications, and is used in many environments to help capture requirements in a clear and unambiguous way.

State machines have plenty of advantages over the use of text for requirements specifications. While text is very flexible and allows for subtle nuance, state machines are precise and unambiguous. It is because of this precision and lack of ambiguity that state machines are so powerful in requirements specifications. In addition, state machines are executable, a characteristic that is invaluable in determining the consistency, correctness, and accuracy of complex sets of requirements. State machines are testable, which allows requirements to be validated before the system is actually designed. Furthermore, tests can be automatically generated from such formal requirements and be applied to the completed system later. Text remains useful—there is no convenient or obvious way to explain "Why?" with state machines alone—but state machines are even more useful than text for the specification of requirements.

Because use cases group requirements into coherent clumps, state machines will be applied a use case at a time. When the system, while executing a use case, primarily waits for incoming events and then responds to them, a state machine specification for the use case is appropriate. These events may be synchronous, asynchronous or time-based, but whenever the system waits for a set of events and executes actions in response, a state machine can easily specify the required behavior of the use case. For use cases that simply run until they are done—that is, they execute algorithms—activity diagrams are usually more appropriate. In the next section, we will see how activity diagrams can also be used to specify behavior.

When modeling a use-case state machine, there are some simple guidelines to follow:

- Incoming messages or commands become events (possibly with data attached as parameters) in the state machine

- Outgoing messages or commands become actions in the state machine

- Underlying technology and design should not be expressed in this state machine because the purpose of a use-case state machine is to specify the required behavior, not the internal design

- Scenarios for the use case are nothing more (or less) than sets of transition "paths" through the state machine

  - Every scenario for the use case should represent transition paths of the use-case state machine

  - For minimal coverage, every transition should be represented in at least one scenario

  - As mentioned before, use-case scenarios form the basis for the validation test suite of the final system product

Scenarios are a very useful, operational view of the system behavior. However, scenarios are only "partially constructive," meaning that they only tell part of the story. A (usually large) set of scenarios is required to show the complete behavior of the use case. A state machine, on the other hand, is "fully constructive" and specifies all of the behavior for the use case in one place. If necessary, these state machines can be decomposed in the standard ways, but whether a nested state is shown on one diagram or decomposed on another, it is still logically a singular view of the behavior.

Problem 3.10 Specification View: Capturing Complex Requirements

Statecharts are a great way to formally capture a behavioral specification for a system executing a use case. In the problem, you must create two statecharts. The first is for the "Evening Low Volume Mode" use case. Be sure to capture the behavioral requirements—the primary road should flash yellow, and the secondary road should flash red. Flashing should occur at 0.5 Hz, with an ON duty cycle of 75%. Also add an initial delay when entering the mode so that any traffic in the intersection has time to clear before engaging in the flashing control.

The second statechart is more complex. You are to create a statechart for fixed cycle time mode. I recommend you do this in three steps.[4] First capture the behavior ignoring turn lanes and pedestrians. Once you get that part correct, then add turn lanes. Remember that there are two different turn-lane modes, one in which the turn lanes complete before any straight traffic can go (SIM, or "simultaneous" mode) and another where the turn lane is green along with the straight traffic going in the same direction (SEQ, or "sequential" mode). A complicating factor is that you must be able to detect when a turn lane detection event has occurred and remember it until it

---

[4]   Breaking down a problem into small chunks and validating each chunk through execution is part of the "nanocycle" of the Harmony (formerly known as ROPES) process. This approach *greatly* simplifies the task of getting the end-product working correctly.

is time to deal with it—and then you can forget it. Use the Latch State Pattern from [2]: create an and-state for each turn lane that has two states: NoTurnRequested, and TurnRequested. When a turn vehicle request occurs for a given turn lane, transition to the TurnRequested state. Once that request has been satisfied (the turn light turns to Red, for example), transition back to the NoTurnRequested state.

Once turn lanes have been properly handled, add pedestrian lanes.

Hint: initially, don't try to create an optimal statechart with a minimum number of states. Once you have the behavior modeled correctly, you can then attempt optimization, if desired. Your goal here is to aim for correctness and simplicity, even if there is redundancy in the state machine. This will not impact the design of the system at all because at this point we are just specifying the requirements. Therefore, if you're relatively new to statecharts, create two or-states, one for the SIM mode of operation and another for the SEQ mode of operation. Then, in each, detail out the behavior of the through traffic. Then add the management of the pedestrian traffic using the Latch State Pattern.

Problem 3.11 Operational to Specification View: Capturing Operational Contracts

In this last requirements analysis problem, we will take one of the Coyote UAV use cases, Perform Optical Surveillance, and walk through the way the Harmony process® captures requirements. This approach has proven quite useful with large-scale development projects that include a separate systems engineering team (or at least, a distinct system engineering effort), followed by an architectural specification and finally by a decomposition into the various engineering disciplines of electronic, software, mechanical, and even chemical design. In software-only projects, particularly ones in which a high degree of agility is important, a simpler approach may be used. However, in this example, we are going to assume that the systems engineering team will be specifying the detailed requirements and systems-level architecture first, and then, at the subsystem-level, decomposing the subsystems into electronics, software, mechanical, and chemical parts, assigning specific requirements to each. Before we get to the problem per se, let's discuss Harmony's requirement capture process.

The Harmony Requirements Capture Process

The overall Harmony process is shown in Figure 3.2. It is a kind of "V-Process" in which requirements analysis and overall system architecture is specified up front, followed by a set of analysis-design-implementation-validation spirals (the latter will be discussed in more detail in later chapters).

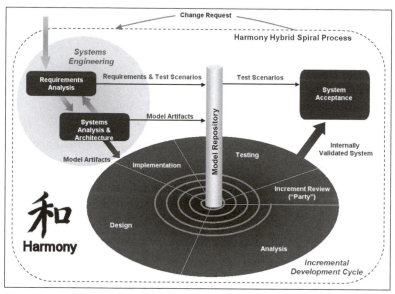

Figure 3.2 The Harmony process overview

In this chapter on requirements analysis, we will naturally be focusing on the requirements capture and specification parts, in the systems engineering part of the overall process. The workflow for the systems engineering parts is shown in Figure 3.3.

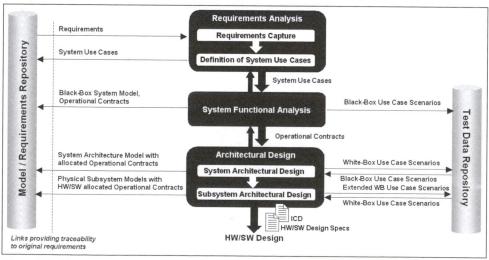

Figure 3.3 Harmony systems engineering workflow

For the CUAV, we have already defined a set of system use cases. In this exercise we will perform the black-box and white-box use-case analysis of one of these. In the "black-box" step of system functional analysis, we will determine the operational contracts into which the system enters when performing the use case. The operational contracts are defined to be the set of services—provided or required by the system—that are used in the operational execution of the system in its environment. This will include the messages sent, the data associated with those messages, and the pre- and post-conditions of the messages. In the next step, we will define subsystems for the CUAV and, finally, in the "white-box" analysis, we will map these operational contracts to the identified subsystems.

A Note on Notation

It is perfectly permissible to use a state machine to specify a use case, and that is what we have done here. However, it is also useful to begin to specify interfaces and that is something that cannot be shown directly on a use-case diagram. For this reason, we will represent the use case with a class. As a class, I can add ports to the use case and specify with precision the interfaces (including the services and data) that define the contracts that the ports support. This will also be useful later when we add internal structure to the collaboration of elements realizing that use case—we can simply nest the object roles within the use-case class as parts. To show clearly that the class is representing a use case, I will preface the name of the class with "uc." I will also represent actors as classes, because the UML lacks a notation to show these as instances or to show the interfaces that they might require or provide—a crucial consideration for systems engineering. Actor class names will be prefaced with "a". For example, Figure 3.4 is a standard use-case diagram that is represented

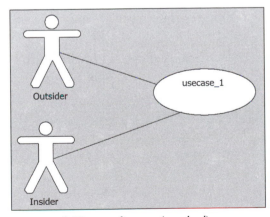

Figure 3.4 Use-case diagram (standard)

using classes in Figure 3.5. The latter figure is more explicit than the former with respect to connection points and interfaces.

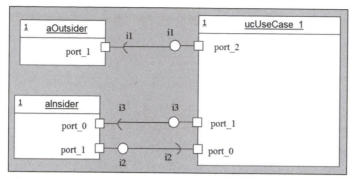

Figure 3.5 Use-case diagram (class notation)

The Rhapsody tool provides the notion of a block, which is nothing more than a singleton object with an implicit class type. Blocks can be converted to normal objects easily, but many systems engineers do not wish to be distracted from their task of capturing requirements with the detailed semantics of objects and classes. If desired, an object or an object role can be used wherever we use the term "block."

Coyote Flight Requirements To Operational Contracts

The basic approach for specifying operational contracts is shown below:

Table 3.1 Mode 2 parameters

Step	Task	Work Product	Comment
1	Define use-case model context	Use-case diagram Structure diagram per use case	Ports are defined but "empty" at this stage
2	Identify system-level operational contracts in the black-box use-case scenarios	"Black-box" sequence diagrams; System-level operational contracts	Lifelines can be actors or use-case blocks
3	Define use-case functional flow	Use-case black-box activity diagram	This task may run in parallel to #2

Table 3.1 Mode 2 parameters (continued)

Step	Task	Work Product	Comment
4	Realize messages as operations	Populated black-box structure diagram	It is easiest not to populate the model with the messages until the sequence diagram is "stable" and then populate it with "auto realize" feature
5	Define interfaces; Allocate messages to interfaces		
6	Define system-level state behavior	Use-case statechart diagram (per use case)	Derived from black-box use-case scenarios
7	Verify and validate black-box use-case model through model execution	Validated system-level use-case model	

For the CUAV use case "Perform Optical Surveillance" we already have a few scenarios. Note that messages from the actor will become operational contracts defined on the ports while "messages to self"—called "reflexive messages"—will be used to specify the functional decomposition of the required processing done in response to the invocation of the operational contract.

Your job, in this exercise, is to:

1. Draw the block diagram representing the use case, the actor(s), and the interfaces between them.

2. The messages that come from the actor(s) must be collected into provided interfaces on the appropriate port of the use-case block and required interfaces on the actor. Messages sent to the actor(s) must be collected into required interfaces on the use-case block and provided interfaces on the actor port. Messages to the use-case block should have public visibility while reflexive messages should be protected or private.

3. For each service (operation) the system provides, specify pre- and post-conditions and the types and ranges of parameters, if necessary.

4. Construct a use-case activity diagram representing all of the scenarios previously specified for the use case.

5. Define a state machine for the use-case block that is consistent with the set of scenarios.

Arguably, the last item in the list will be the most challenging.

References

[1] Douglass, Bruce Powel, *Real-Time UML, Third Edition: Advances in the UML for Real-Time System*s, Addison-Wesley, 2004.

[2] Douglass, Bruce Powel, *Doing Hard Time: Developing Real-Time Systems with UML, Objects, Frameworks, and Patterns*, Addison-Wesley, 1999.

Systems Architecture

What you will learn:

- **Essential properties of systems architecture**
- **Organizing systems models**
- **Identifying systems architecture**
- **Specifying subsystem interfaces**
- **Mapping requirements to subsystems**

Overview

This chapter is specifically for projects that contain a systems aspect—that is, systems that will ultimately be realized using more than a single engineering discipline. It is common, for example, for systems to contain software, electronic, mechanical, and even chemical parts. The systems architecture identifies requirements of the system as a whole and the architecture of the system constructed of subsystems that are themselves to be implemented with some combination of engineering disciplines. In the previous chapter we discussed capturing of systems requirements. In this chapter we will focus on the specification of a systems-level architecture.

Many projects are primarily software-oriented and don't require effort defining the systems architecture, although significant effort should be spent in specifying the software architecture (the subject of Chapter 6). However, even for software-only projects, this chapter may be highly useful as it introduces the important aspects of architecture and focuses on one of these—the identification of subsystems, with the concomitant facets of decomposition of system use cases into subsystem use cases, and the specification of interfaces and interactions among the subsystems.

Let us review what we mean by the term *architecture*. Architecture encompasses the large-scale design decisions that affect most or all of the system. Note that architecture is a part of design. In the Harmony process, analysis is all about capturing the essential aspects of the system—the properties and characteristics of the system that are essential for correctness. Design, on the other hand, focuses on optimization of the analysis model against what are collectively termed the design criteria—the set of aspects of the system against which different designs and technological choices are measured, evaluated, and, ultimately, selected. Design criteria may refer to system performance (such as worst-case execution time, bandwidth, or throughput), run-time usage of system resources (such as memory), design-time "goodness" metrics (such as complexity or encapsulation), design properties (such as maintainability, reusability, or portability) or even project properties (such as work effort required).

The Harmony process identifies three levels of design. Architectural design attempts to optimize the entire system scope at a gross level with a coherent set of architectural choices. Mechanistic design seeks to optimize a collaboration of objects working together to realize a system-level capability (e.g., a single use case). Detailed design optimizes individual objects. We'll focus on just architecture here since that is the perspective of system engineering.

The Harmony process identifies five key aspects of architecture, highlighted in Figure 4.1.

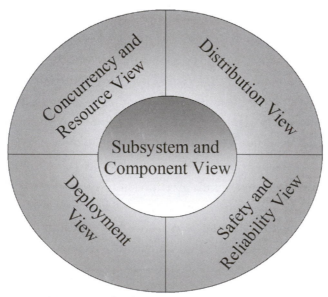

Figure 4.1 Aspects of architecture

The concurrency and resource view of the architecture identifies the concurrency aspects of the system and how resources are shared among those concurrent units. The distribution view specifies how the objects will be distributed across multiple address spaces (e.g., CPUs) and how they will communicate and collaborate efficiently and reliably. The safety and reliability view of the architecture concerns itself with the identification, isolation, and correction of faults during system execution. The deployment view identifies which aspects of the system shall be implemented in the various engineering disciplines and how they will interact. Finally, the subsystem and component architecture view concerns itself with the specification of the largest scale pieces of the overall system, how the system functionality maps into those structures, and how those elements interact in the large scale. It is this last aspect of architecture that is the primary focus of system engineering, although the other aspects may be considered as well.

What we mean by a system architecture is the identification of the strategic design decisions that affect most or all of the system from the systems point of view. The systems point of view is "above" software, electronic, or mechanical engineering. Thus, the systems architecture will focus on the specification of the set of subsystems into which the system will be decomposed, the allocation of requirements and functionality to those subsystems, and the interfaces between those subsystems. At the end of the systems architecture activities, portions of the systems model are handed off to the subsystem teams, where each subsystem is then decomposed into the engineering disciplines and more detailed analysis and design work begins.

To this end, the systems architectural definition will focus almost exclusively on the subsystem and component view of the architecture, although it may have to "drill down" into the other views from time to time.

Problem 4.1 Organizing the Systems Model

Organizing the model is something not usually considered until the team runs into a problem managing the burgeoning complexity of a system under construction. Reorganizing a model at that time requires a nontrivial and (almost always) unscheduled effort. Projects that have a systems team performing systems analysis are usually complex enough to warrant some early consideration of how you would like to manage the model.

There are many ways to organize models and, while we will focus on only one here, there are viable alternatives. As with most things, though, there are many more bad ways to organize models than good ones. Some points to consider include:

- Large project or small?

- One model or many?

- Single product or family of products?

- How will common parts be shared among team members?

- How will the architectural aspects be made available to the team?

- How will the architectural decisions be enforced among the team?

- Will the team be co-located or distributed?

- How will we minimize model management overhead?

- What principles will be used to group elements together?

- How will configuration management be used?

One of the first considerations is whether to have one large model shared among the team or separate models. A single model has the advantage of simplicity of management—you only have one entity to manage (even if it has subparts). On the other hand, a single model takes longer to load—a real concern if you have 50 or more members on a team—and finding things to facilitate sharing and collaboration in a large monolithic model can be difficult. Multiple models have the advantage that system complexity can be divided across many different models and each model is smaller, more manageable, and takes less time to load than a larger single model. On the other hand, significant thought should be put into what the submodels should be, what criteria should be used to locate model elements in the various models, and how the models will be shared across multiple stakeholders.

As a recommendation, for teams of 15 or fewer members, a single model may be a good choice; for teams of 20 or more, multiple collaborative models are probably a better choice. Of course, the properties of the system under development impact the decision. If the model is linearly separable (i.e., able to be broken into a number of more-or-less independent pieces), management of multiple models is easy. If the system is not linearly separable, it may be much more difficult to create multiple models. If, in order to work on one model one must import all the other models as well, breaking the system up into multiple models won't help you.

The primary reasons to break up large models are to 1) decrease load/save times and 2) provide smaller but sufficient models to teams who have narrow focus. It is this latter concern that we will address here. The typical stakeholders for the system model are shown in Table 4.1.

Table 4.1 Model stakeholders

Group	Purpose	Scope of Concern
Systems engineers	• Construct a discipline-independent model of requirements and architecture • Map requirements and functionality to subsystems • Specify interfaces between subsystems	• System requirements • System architecture
Subsystem team	• Create a deployment architecture for the specific subsystem • Create software and hardware specifications (submodels)	• Single subsystem • Mapping of subsystem requirements to engineering disciplines
Engineers (SW, mechanical, electronic, chemical)	• Perform analysis and design of a specific subsystem within their discipline	• Model elements within their discipline for a single subsystem
Design architect	• Oversee architectural design for the set of subsystems	• Common architectural model for entire system • Discipline-related architecture for each subsystem
Testers	• Subsystem-level testing • Integration testing • Validation testing	• Subsystem—normally done within the subsystem team • Integration—normally done with system-level scope • Validation—normally done at system-level scope

For small-scale systems, we recommend the following model organization:

- Systems area
  - Systems requirements
    - Operational requirements (use cases)
    - Non-operational requirements
    - System test vectors
  - Systems architecture (including deployment)
- System builds area (for incremental construction of the system)
  - Build x (one per build)
- Common area (for shared types and classes)
- Collaboration x (one per system use case)

The above organizational scheme has the advantage of simplicity, but lacks scalability to large team sizes and large problems. For moderate to large-scale systems, in order to satisfy the needs of the various stakeholders, we recommend the following, more elaborate organization:

- Systems area
  - Systems requirements
    - Operational requirements (use cases)
      - Use case x (for each use case)
        - Black box
        - White box
    - Nonoperational requirements
    - System test vectors
  - Systems architecture
- System builds area (for incremental construction of the system)
  - Build x (one per build)
- Common area (for shared elements)
  - Subsystem interfaces
  - Shared domains
- Subsystem area x (for each subsystem)
  - Subsystem requirements
    - Subsystem operational requirements (use cases)
      - Use case x (for each subsystem use case)
    - Subsystem nonoperational requirements
    - Subsystem test vectors

 – Subsystem architecture
   ▪ Subsystem deployment
   ▪ Interdisciplinary interfaces (e.g., sw-electronics, electronics-mechanics)
 – Collaboration x (one per subsystem use case)

This organization can be used for models of moderate to very large scale. For more moderate-scale projects, the black box/white box use case packages can be omitted and the set of use cases can be stored together in a single package. For small models, a single requirements package that holds every requirements aspect (operational, nonoperational, requirements elements, use cases, constraints, etc.) may be sufficient. The organizational scheme can be used within a single model, but it also suggests division points for multiple models. For large systems projects, we recommend the following subject areas be broken out into separate models:

- Systems model
- Common area model
- Subsystem models (one per subsystem)

In UML, the unit of model organization is the package. A package is a model element that contains and provides a namespace for other model elements, some of which may be (nested) packages themselves. Model organization can be shown in either a package diagram (a class diagram whose purpose is to show the organization of packages in a model or set of models[1]) or in a browser (tree) view.

For this first problem,

1. Create a single model for the Roadrunner Traffic Control System organized as above. Since this is a more moderate scale project, use a single package for all the use cases.

2. Create a set of models for the Coyote UAV using the suggested large-scale organization discussed above. Since this is a large-scale model, use separate packages for the system-level use cases, with nested black box/white box packages.

Note that we haven't yet identified the subsystems or domains, so you can leave the subsystem area of your model empty. We will elaborate this model organization as we identify subsystems and domains later in this chapter.

---

[1]   A model is a kind of package.

Problem 4.2 Subsystem Identification

At this point, we have selected an overall model organization (the "logical architecture") for the system. It may be in one model—as is the case with the Roadrunner Traffic Light System—or it may be a set of interrelated models—as is the case for the CUAV system. This means that we've identified the set of organizational units that exist at design time to help us manage the complexity of the problem we have set about to solve. We need to make a similar set of decisions about how we organize the set of run-time elements as well. Packages, the elements used to organized the models, only exist at design time. Packages are not instantiable elements and serve only to organize the elements of the design.

To organize the elements that exist in the running system (e.g., the objects) we must use instantiable elements, objects of some kind. In the UML, this means instances of classes. The UML defines a couple of kinds of classes that are used for large-scale run-time organization. Specifically, subsystems and components are used for this purpose, but these are basically just structured classes. A structured class is nothing more (or less) than a class that contains internal parts (that are themselves specified by classes). A subsystem or component is basically a large-scale structured class. There are a few technical differences, such as a component associated with an artifact, but these differences are minor. An artifact, for our purposes here, is the embodiment of the run-time entity, such as a .DLL, .EXE, or .LIB file that manifests the object. A subsystem is just a kind of component. Really, it's all just about classes.

While the UML defines the elements *class*, *system*, *component*, and *subsystem*, it really doesn't mandate how they are used. The Harmony process recommends, therefore, a particular set of guidelines that we have found effective:

- A system element is a class that encompasses the largest-scale thing in your project.

- A subsystem element is a class that is the first-level compositional unit of a system element (that is, subsystems are parts of systems).

- A component is a software class that is the first-level compositional unit of a subsystem.

- A task is a structured class, contained within a component, that is active (i.e., owns the root of a thread) and is the fundamental unit of concurrency in the model.

In practice, run-time systems are organized by large-scale objects, and these instances are typed by classes. In very large-scale systems, your project might contain multiple instances of each of these levels. In small systems, you might skip some levels of organizational abstraction altogether.

So what makes up a good set of subsystems? How is functionality decomposed into subsystems? How do subsystems interact? Over the years, I have developed a set of guidelines that I have found useful for subsystem identification:

- A subsystem should be loosely coupled with other subsystems

- Subsystems should enforce encapsulation through the use of ports and contract-driven subsystem design

- A subsystem's internal elements should be more tightly coupled than they are with elements in other subsystems (and often don't use ports)

- A subsystem should contain elements that contribute to a small number of coherent functions

- A subsystem should have a well-specified set of interfaces, consisting of

  - A set of services provided by the subsystem

  - A set of services required from other subsystems

  - Pre- and post conditions for each service

  - Constraints and invariants for each service

- A subsystem should use a common set of hardware to support its functionality

- In a multiprocessor system, a subsystem normally operates entirely on a single processor or names the coherent set of functionality operating on a tightly coupled set of processors

- If multiple design teams are to be employed, a subsystem should be designed by a single team

Of course, these are guidelines and not strict rules—good design is still as much an art as it is an engineering discipline.

Ports are often used in the subsystem architecture. Ports are a design pattern that:

- Allows the explicit delegation of services defined on the subsystem to internal parts

- Enforces encapsulation by not allowing outside clients to have direct knowledge of the internal structure of the structured class

- Explicitly specifies interface contracts

- Introduces some level of extra complexity and has some small performance and memory overhead that cannot always be optimized away

Because of the overhead of port usage, ports are not recommended for use everywhere, but for subsystems and architectural objects, ports provide benefits that usually outweigh their costs.

For this problem, using the guidelines above, identify the subsystem architecture for both the Roadrunner Traffic Light System and the CUAV. This means to identify the subsystem objects (or classes, if you prefer) and draw the subsystem architecture using a structure diagram with the outermost class being the entire system, with internal parts representing the subsystems. Add ports and connect the subsystems together with links as you think appropriate. Understand that these ports and links may change as we do further analysis and design.

The Roadrunner model contains a _System/SystemsArchitecture package.[2] Place the subsystem classes, objects, and diagrams in this package. Note that the CUAV may have multiple levels of subsystem (such as the air vehicle, the ground station, and subsystems within those primary systems). Since the CUAV is spread over multiple models, identify the subsystems in the Systems Engineering model in the Architecture package.

Problem 4.3 Mapping Operational Contracts into Subsystem Architecture

As we have seen, requirements are captured in a number of forms in model-based development projects. One of the key forms is the scenario, a sequenced set of service invocations during the collaboration of a system. A scenario is a path through a use case, in which a particular set of services are executed in a particular sequence to get a specific result. At the highest level ("black box") of abstraction, the entire system becomes a single lifeline on the sequence diagram. However, system-level things do very little work in and of themselves; primarily they organize and orchestrate the behavior of their internal parts to achieve the system's operational goals. This means that a measure of the "goodness" of the subsystem architecture is whether they can collaborate together to achieve the operational goals of the entire system. If they can do so in an efficient way, then the subsystem architecture is "good"; if

---

2   I tend to use the underscore in the _System package name to ensure that it comes first alphabetically, even though in later versions of Rhapsody this isn't strictly necessary.

they cannot, then it "needs improvement." The set of services in a set of scenarios, between two elements at the same level of abstraction, constitutes those elements' "operational contract" and will be used to specify the interfaces on the ports between those elements.

In this exercise, you will be given a use case for each of the two examples and one or more scenarios. Your job will be to take the scenario—and the operational contracts that it implies—and map it down to the subsystem level of abstraction and demonstrate how the subsystems collaborate together to realize the required use-case behavior.

Figure 4.2 shows the use cases for the Roadrunner Traffic Light Control System. In this exercise we will select one use case and one or two scenarios and do the subsystem elaboration ("white box") view, identifying the roles that the subsystems play in the execution of the scenarios.

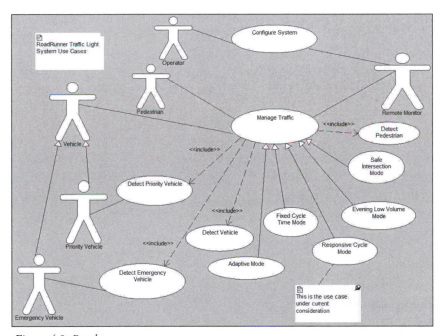

Figure 4.2 Roadrunner use cases

The use case we will consider is "Responsive Cycle Mode." Remember, in this mode, the system senses and responds to vehicles and pedestrians by setting the lights appropriately. In the next figure, Figure 4.3, we see how this use case plays out in one particular case.

The next figure shows a different scenario for the same use case. Just to illustrate a slightly different way to model, in the previous scenario (Figure 4.3) the system

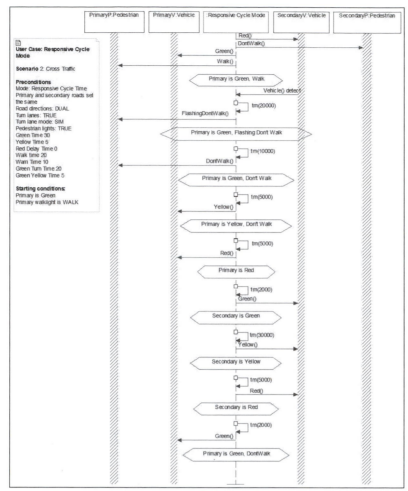

Figure 4.3 Roadrunner Responsive Cycle Mode scenario

use case is shown on the lifeline; in the next figure (Figure 4.4) the system object is used. The difference is immaterial and which you use is a matter of personal choice. If you use the use case as a lifeline, then it represents the system executing that functionality; if you use the system object, then it still represents the system, but in this scenario you're still focused on the functionality in that one use case. Whichever you prefer, these two scenarios are illustrating different ways in which the Responsive Cycle Mode use case can play out. In the former case, a vehicle arrives on the secondary road, causing the traffic light controller to cycle through the lights. In the latter case, a pedestrian signals that they would like to cross the street, causing the traffic light controller to cycle through the lights to allow the pedestrian to traverse the street.

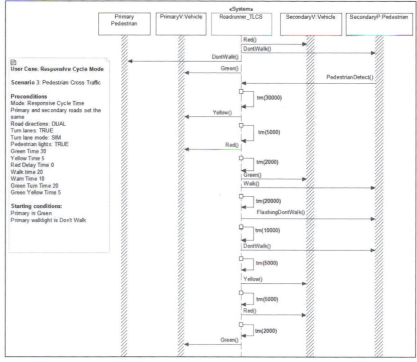

Figure 4.4 Roadrunner Scenario 2

So take these two scenarios and elaborate them to include the subsystems identified for the Roadrunner Traffic Light Control System, showing how the subsystems collaborate to realize the required system-level behavior.

The second part of this exercise is to do the same thing for the CUAV. We want to map the operational contracts identified in the Perform Area Search use case. The use case, and its description, is shown in Figure 4.5. The CUAV is, of course, a large-scale complex system. The Perform Area Search use case includes other use cases as summarized in Figure 4.6; since Perform Area Search is a kind of Execute Mission use case, it includes (indirectly) navigating and flying the UAV, managing the datalink, acquiring and processing surveillance data, and so on. So when we model the operational contracts of the complete Perform Area Search, a great deal of the CUAV behavior is ultimately included. We will simplify it somewhat here to make the exercise more tractable, but we will include enough to give you a flavor of what's involved. The beginning student shouldn't be too discouraged at the complexity of the system—after all, aircraft requirements and systems analysis usually take place over several years by a set of highly specialized engineers.

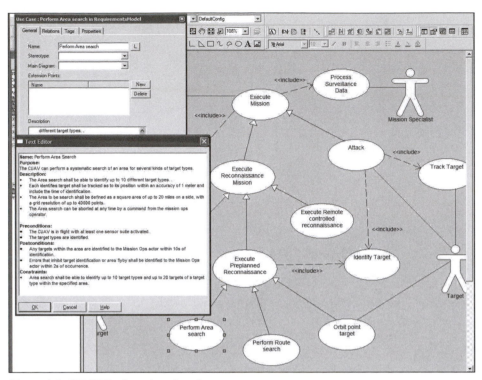

Figure 4.5 CUAV Perform Area Search use case

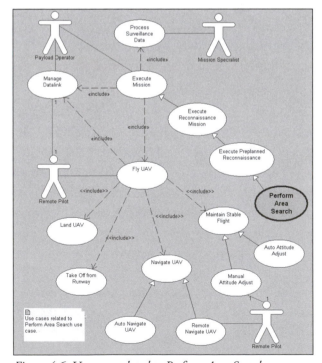

Figure 4.6 Use cases related to Perform Area Search

Because of the length of the scenario (and to illustrate the technique), the scenario is broken across multiple sequence diagrams. The first is, in some sense, the "master," or high-level view. It is broken up into three subdiagrams, indicated with the interaction references. Each of these refers to a separate diagram focusing on some part of the overall scenario. In fact, this high-level sequence diagram is decomposed along the same lines as the included use cases, and this is not accidental.

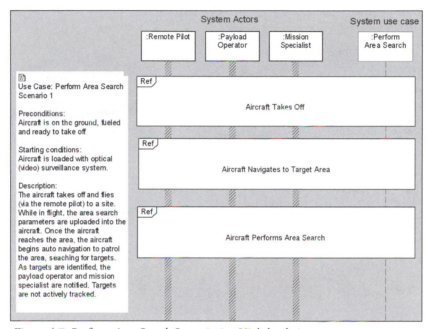

Figure 4.7 Perform Area Search Scenario 1—High-level view

Figure 4.8 shows the elaboration of the aircraft taking off. Note that this is simplified to show the set of commands used by the pilot to control the aircraft and the set of the responses and status messages from the aircraft to the pilot, but not the precise order. This shows a useful way in which this kind of system—which is continuously controlled by the pilot, but that control is mediated by a set of discrete commands and messages—can be modeled. Note the use of the par (aka "parallel") operator. In this case, it means that these two sets of interactions are independent and can be interleaved in any fashion. The upper section depicts commands from the pilot and the lower set depicts status messages from the aircraft.

Once the aircraft takes off, it must navigate to the desired location. Of course, the remote pilot is, in this case, flying the aircraft remotely. It would be onerous to capture the hundreds or possibly thousands of individual messages and commands

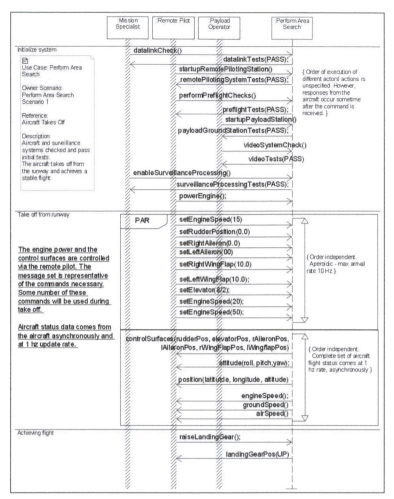

Figure 4.8 Perform Area Search Scenario 1—Aircraft takes off

issued as a result of flight adjustments to the remote controls. As in the previous sequence diagram, we have satisfied ourselves with the set of commands that can be issued (see Figure 4.9). We've added a constraint to show that the commands from the pilot are order independent—that is, he can adjust any control at any time and in any order, but the commands are sent to the aircraft no more frequently than 10 commands/second.

In the next figure, Figure 4.10, the aircraft is commanded to search a specific area. As a part of this, the aircraft makes a grid of the rectangular area searched, and flies over the area in each of the grid's "columns" search for targets. When a target is found, its location is noted and transmitted to the ground station. The aircraft also maintains a list of identified targets. While doing an area search, the aircraft navigates via its internal autopilot.

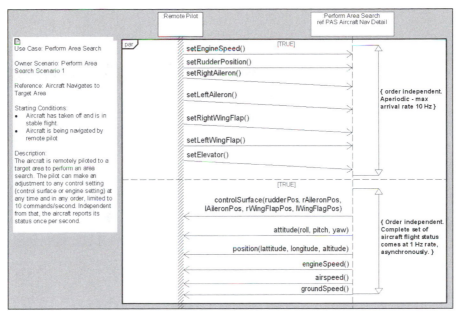

Figure 4.9 Perform Area Search Scenario 1—Remote navigation

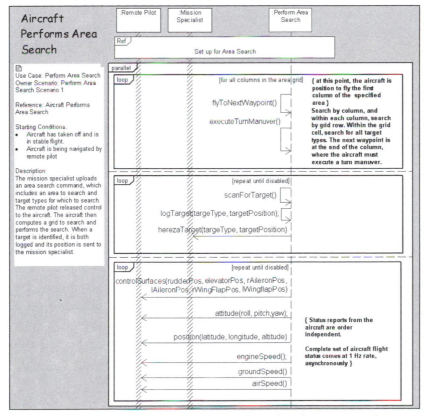

Figure 4.10 Perform Area Search Scenario 1—Aircraft performs area search

The nested scenario in Figure 4.10 itself has a nested scenario for the set-up of the area search. This is shown in Figure 4.11.

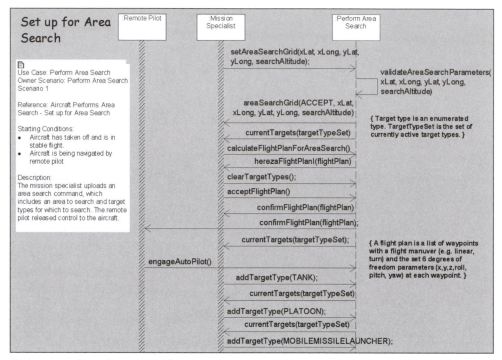

Figure 4.11 Setup for area search

For the second part of this exercise, map the operational contracts identified in the use case Perform Area Search into the Coyote system and subsystem architecture. Remember that the Coyote UAV is a large-scale system. We are working with only a single scenario of a single use case. In a real system development, there would be an average of a dozen or more scenarios for each use case and this process would be repeated for all scenarios and all use cases. Once several primary scenarios for a use case are completed, additional "sunny day" scenarios may add only a few, if any, operational contracts. The "rainy day" scenarios that model what happens when things go wrong add more services and behavior. In any case, it is an important exercise to go through to ensure that all required services are provided by some element in the system, the system's overall behavior is coherent, and the system architecture meets all the requirements.

Problem 4.4 Identifying Subsystem Use Cases

Large projects are almost always constructed by multiple teams. The most common way to organize such teams is by allocating subsystems to different teams. A subsystem is a large-scale architectural piece of the physical system. It may be composed of software alone or be a combination of multiple engineering disciplines, typically software, electronics, mechanical, and chemical. These teams may or may not be colocated but they need two distinct kinds of information to proceed effectively.

First, they need to know the requirements for their particular subsystem. If I'm on a team constructing a power subsystem for a spacecraft, I need to understand what capabilities are expected of my subsystem by its clients—the system, the system actors, and peer subsystems. We organize the requirements of systems into use cases, and it seems reasonable to do the same for subsystems.

Secondly, we need to know the operational context into which the subsystem must fit: what are the clients of my subsystem's services and what servers are available for my subsystem to use? I need to know the interfaces that my subsystem provides to those clients, including the pre- and postconditions, data contents and types, quality of service constraints, and the allowable set of sequences of those services. I need the same information for interfaces that my subsystem requires.

These two kinds of information are provided from two sources, the systems requirements model and the systems architecture. These two kinds of information are clearly codependent. A different system architectural structuring will result in a different set of use cases for the subsystems.

So the question is, "How is the identification of the subsystem use cases actually done?" As mentioned above, it requires two sources of information, the system use case model and the system architecture. The job is then to decompose the system use cases into a set of subsystem use cases that are then wholly met by individual subsystems. This is semantically accomplished with the «include» dependency. The basic algorithm for the generation of the subsystem use cases is:

- For each system use case
  - Identify clusters of requirements that are met by a single subsystem
  - Create a subsystem-level use case to organize that cluster
  - Add the «include» dependency to link the subsystem use case to its parent system-level use case
  - Allocate the use case to the subsystem

      – Move the allocated subsystem use case to the requirements area for its specific subsystem

At the end of this process, each subsystem has a set of use cases that capture its requirements.

Every operational and operational quality of service requirement must be represented in at least one subsystem use case; some requirements will be represented in all or most of the subsystems. Other requirements must be decomposed into derived requirements, each of which maps into a single use case on a single subsystem. The idea is that the subsystem requirements model represents a complete set of requirements for that particular subsystem. The subsystem team can then go construct their system more-or-less independently, with assurance that their subsystem will fit into the system architecture and function appropriately. Traceability is achieved because the links between the system and subsystem use cases are traceable dependencies. Explicit links to requirements traceability tools, such as DOORS, are commonly added as well.

This effort results in new use-case diagrams focused at a more detailed level of abstraction. In general, you want to create a new use-case diagram for each decomposed system-level use case. In addition, you want to create one or more new use-case diagrams for each subsystem. The missions of these two kinds of use-case diagrams are different. The former shows the tracing of the parent-system use-cases to their decomposed subsystem use cases. These diagrams will be kept in the system

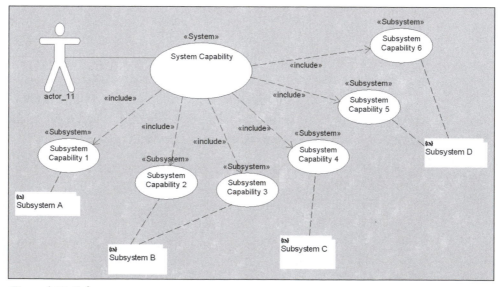

Figure 4.12 Subsystem use cases

requirements area of the model. The latter shows the set of use cases for the subsystem, and will be located in the subsystem area of the model. The subsystem-level use cases themselves will be specified in their associated subsystem package in the model. Eventually, the subsystem area of the model will be passed off to the subsystem teams and they will elaborate this package in their own design models.

For the first part of this exercise, use the subsystem architecture for the Roadrunner Traffic Light Control system. Take the Configure System and Detect Vehicle use cases, which we have specified in some detail earlier. Figure 4.13 shows this use case in the use-case model.

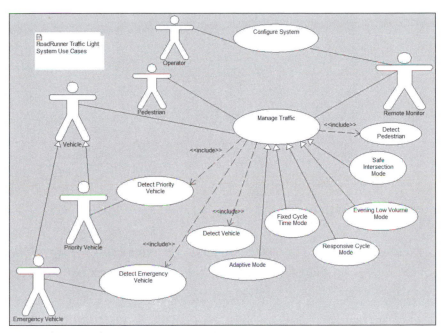

Figure 4.13 Roadrunner use cases

The Roadrunner architecture is shown in Figure 4.14. The functionality (and requirements) represented by the Configure System use case must be mapped into a set of use cases met by the set of subsystems. This system-level use case may not impose any requirements on some subsystems or may result in multiple use cases on some of the subsystems. It is exceedingly rare for a system-level use case to map entirely onto a single subsystem, however.

To complete this portion of the exercise, create new use-case diagrams for the Configure System and the Detect Vehicle use cases; in these new diagrams, decompose the system use case into a set of subsystem-level use cases. Once done for each subsystem-level use case, create a use-case diagram for the IntersectionController subsystem with all of its identified use cases.

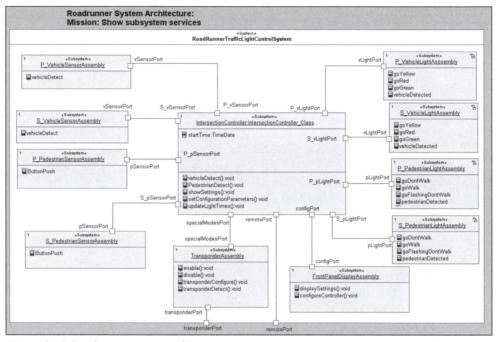

Figure 4.14 Roadrunner system architecture

The second part of this exercise is to do the same thing for two of the use cases for the CUAV—Manual Attitude Adjust and Perform Area Search use cases. These two use cases are highlighted in Figure 4.15 and Figure 4.16.

The first of these is the simpler of the two. The capability described by the Manual Attitude Adjust means that a remote pilot in the ground station can control the attitude (roll, pitch, and yaw) of the aircraft. To do this, the pilot will need to monitor that information and have it displayed in some meaningful way and then be able to control these aspects of the aircraft via the communications link. This will have impacts on many subsystems, both in the ground station and in the aircraft. The overall system architecture is shown in Figure 4.17. The two primary systems are shown in the next two figures. The aircraft itself is detailed in Figure 4.18. This shows the air vehicle subsystems and their interconnections. The ground station is shown in Figure 4.19. This figure is less complex, but shows that there is a set of manned control stations, each of which contains a UAV Control Station and a UAV Monitoring Station.

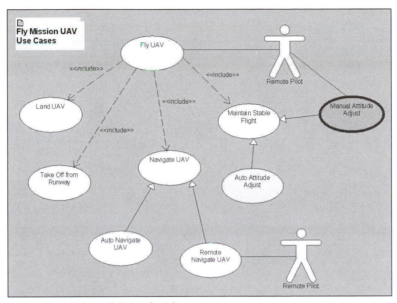

Figure 4.15 Manual Attitude Adjust use case

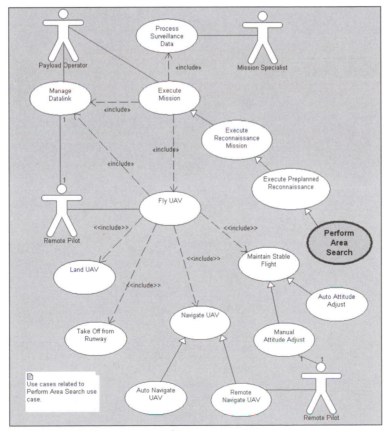

Figure 4.16 Perform Area Search use case

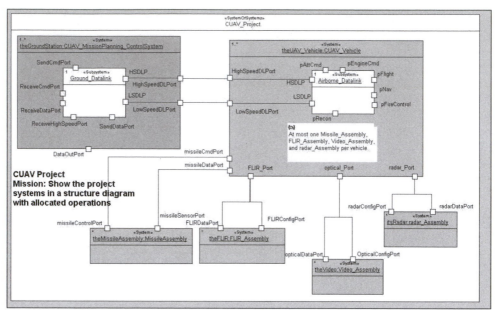

Figure 4.17 CUAV System architecture

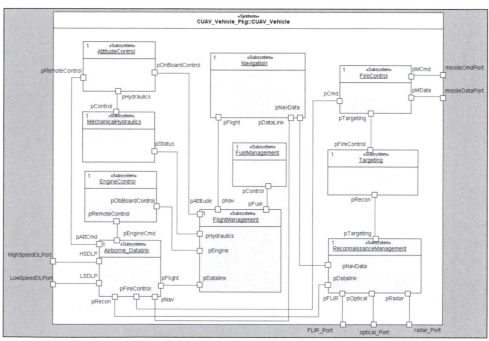

Figure 4.18 CUAV Aircraft architecture

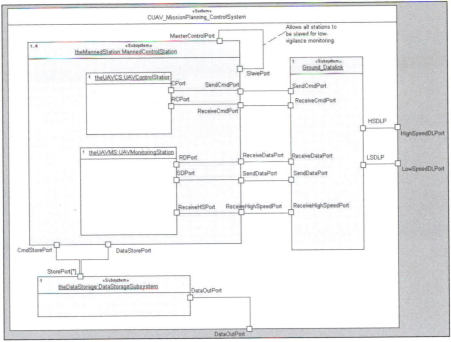

Figure 4.19 CUAV Ground station architecture

The first part of the problem of identifying subsystem use cases is to map the Manual Attitude Adjust use case to the set of subsystems identified here. The second part of the problem is to map the Perform Area Search use case to the same subsystem architecture.

Looking Ahead

In this chapter we did exercises to create the systems architecture, including the specification of the subsystems and their requirements. The next step is to analyze the use cases and construct what used to be called the essential model of the system, but what is now more commonly called the platform-independent model. This model constructs a set of objects, complete with their classes, relations, and behaviors, that work together to realize the requirements of the use case. The Harmony process calls this effort *object analysis*. That is the subject of the next chapter.

Object Analysis

What you will learn:
- How to do object analysis
- Underline the noun strategy
- Identify causal agents strategy
- Identify services strategy
- Identify messages strategy
- Identify real-world items strategy
- Identify physical devices strategy
- Identify key concepts strategy
- Identify transactions strategy
- Identify persistent information strategy
- Identify visual elements strategy
- Identify control elements strategy
- Apply scenarios strategy

Overview

The purpose of object analysis is to:
- identify the objects and classes essential to realizing the use cases
- identify relations among the essential objects and classes
- identify attributes and allocate them to classes
- identify operations and allocate them to classes
- specify behavior of reactive classes with state machines
- specify operation behavior with activity diagrams
- validate the correctness of the analysis model

Since the requirements are organized into coherent use cases, object analysis is normally applied one use case at a time, resulting in an object collaboration that realizes all of the functional requirements of the use case. This collaboration is unlikely to be optimally efficient, but that is where object design comes in—the optimization of the object analysis model. But first, we must identify the essential objects and classes, the ones that absolutely must be there for the system to be correct.

Object analysis is a crucial phase of the Harmony spiral model.[1] In each iteration through the spiral process, a number of use cases are elaborated into an executable, validated prototype. Figure 5.1 uses an activity diagram to depict the workflow for object analysis.

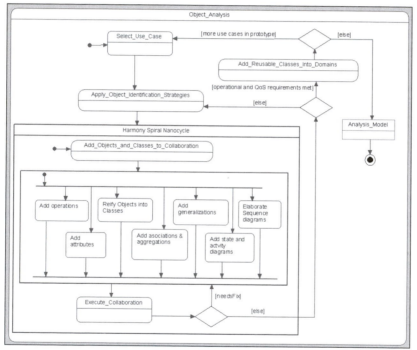

Figure 5.1 Object analysis workflow

For each use case in the prototype build, we identify objects and classes with object identification strategies, and refine the models by adding details such as operations, attributes, and various relations. The key to effectively constructing the collaboration to realize the use case is to do continuous validation all the way through its construction. Rather than create a collaboration of 50 objects and only then begin to execute, it is far more effective to begin execution as soon as a single object is added and to repeat the execution with each small incremental change. In this way,

---

[1]   Chapter 2 discusses the Harmony process workflows in more detail.

we can identify defects immediately and remove them just as quickly. Rhapsody provides a very powerful environment with its model execution and debug facilities that can greatly enhance this "extreme" or "agile" modeling approach.

In object analysis, we discover (more than invent) the essential properties of the system and represent them as objects, classes, attributes, operations, states, transitions, and relations. By "essential," I mean that the collaboration is wrong if it doesn't include that element. For example, if I am modeling a "Make Deposit" use case for a banking system, I would expect to see objects representing the customer, the bank account, the value held within the bank account, and the transaction itself. If those things are not represented, then the collaboration doesn't properly represent the requirements in the use case. If we have a use case "Heat Food in Microwave Oven," then I would expect to see a door, door sensor, microwave emitter, a mechanism to vary the microwave intensity, and timer (actual or implied)—or it's not a microwave oven cooking food, but something else. In the object-analysis step in the Harmony process, we identify these essential properties and how they are related to each other. And we do this in an executable fashion so that we can, at any time, validate the portion of the object model constructed thus far.

Key Strategies for Object Identification

There are many ways to identify the objects within a collaboration. Table 5.1 outlines what I have found to be the most effective of these object-identification strategies. These strategies are a key technique in the development of software with the Harmony process.

Table 5.1 Object discovery strategies

Strategy	Description
Underline the noun	Used to gain a first-cut object list, the analyst underlines each noun or noun phrase in the problem statement and evaluates it as a potential object, class, or attribute.
Identify causal agents	Identify the sources of actions, events, and messages; includes the coordinators of actions.
Identify services (passive contributors)	Identify the targets of actions, events, and messages as well as entities that passively provide services when requested.
Identify messages and information flow	Messages must have an object that sends them and an object that receives them as well as, possibly, other objects that process the information contained in the messages.

Table 5.1 Object discovery strategies (continued)

Strategy	Description
Identify real-world items	Real-world items are entities that exist in the real world, but are not necessarily electronic devices. Examples include objects such as respiratory gases, air pressures, forces, anatomical organs, patient information, chemicals, vats, etc.
Identify physical devices	Physical devices include the sensors and actuators provided by the system as well as the electronic devices they monitor or control. In the internal architecture, they are processors or ancillary electronic "widgets." Note: this is a special kind of "Identify real-world items."
Identify key concepts	Key concepts may be modeled as objects. Bank accounts exist only conceptually, but are important objects in a banking domain. Frequency bins for an on-line autocorrelator may also be objects. This strategy is an antipode to the "identify real-world items" strategy.
Identify transactions	Transactions are finite instances of interactions between objects that persist for some significant period of time. Examples include bus messages and queued data. This may be done with ACID (atomic, complete, isolated and durable) transactions or any kind of transactions, state-based or otherwise.
Identify persistent information	Information that must persist for significant periods of time may be objects or attributes. This persistence may extend beyond the power cycling of the device.
Identify visual elements	User-interface elements that display data are objects within the user-interface domain such as windows, buttons, scroll bars, menus, histograms, waveforms, icons, bitmaps, and fonts.
Identify control elements	Control elements are objects that provide the interface for the user (or some external device) to control system behavior.
Apply scenarios	Walk through scenarios using the identified objects. Missing objects will become apparent when required actions cannot be achieved with existing objects and relations.

We will briefly discuss all these strategies, but note that the analyst need not use them all on any specific project. These approaches are not orthogonal and the objects they will find will overlap to some degree. The best way to use these strategies is to find the two or three that work best for you and the problem at hand and apply this selected set of strategies to find all of the objects and classes in the analysis collaboration.

Underline the Noun Strategy

The first strategy works directly with the written problem or mission statement. Underline each noun or noun phrase in the statement and treat it as a potential object. Objects identified in this way can be put into different categories:

- Objects of interest
- Classes of interest
- Actors
- Uninteresting objects
- Attributes of objects
- Events
- Synonyms for any of the above

Identify the Causal Agents

For every effect, there must be an element that is the cause. This strategy looks to identify the objects that cause things to happen—i.e., the ones that are behaviorally active ones. These are objects which:

- Produce or control actions
- Produce or analyze data
- Provide interfaces to people or devices
- Store information
- Provide services to people or devices
- Contain more fundamental objects as parts

A causal object is an object that autonomously performs actions, coordinates the activities of component parts, or generates events. Whenever there is some initiating action, an object somewhere in the system must provide that service.

Identify Services (Passive Contributors or Server Objects)

In an object-oriented analysis, all (or at least the great majority of) services will be provided by objects. They may provide passive control (that is, they do what is requested of them when it is requested), data storage, or both. A simple switch is a passive control object. It provides a service to the causal objects (it turns the light on or off upon request), but does not initiate actions by itself. Passive objects are also known as servers because they provide services to client objects. Incidentally, servers almost always are at the receiving end of unidirectional associations from their clients; that is, clients know about the servers, but the servers do not know their clients.

Identify Messages and Information Flows

For each message, there is an object that sends it, an object that receives it and, potentially, other objects that process it. Messages correspond to information and control flows and are realized by either operation calls or event receptions. In either case, for every message, there is at least a sender object and a receiver object. In addition, the message itself is often an object. Certainly that is true for remote communications with network packets and datagrams. It is also true when the message must be remembered or processed in some way (see the Identify Transactions strategy, below).

Identify Real-World Items

Embedded systems need to model the information or behavior of real-world objects even though they are not part of the system per se. An anesthesia system must model the relevant properties of patients (name, weight, condition, billing information, etc.), even though customers are clearly outside the anesthesia system. A tracking system would typically model relevant aspects of the things that it tracks, such as combat ID, target type, location, velocity and acceleration. These internal representations of real-world elements are normally not simulations of those systems but representations of information it is important for the system to represent. However, in some cases, such as simulation systems, it is crucial to model the physics of such systems as well, such as airflow in aerodynamic modeling.

Identify Physical Devices

Real-time systems interact with their environment using sensors and actuators in a hardware domain. The system controls and monitors such physical devices inside and outside the system and these objects are modeled as objects. Devices must also be configured, calibrated, enabled, and controlled so that they can provide services to the system. When device information and state must be maintained or services invoked, the devices are modeled as objects. Objects representing physical devices almost always represent the interfaces to those devices and serve as a means of encapsulating the invocation of services from them.

Identify Key Concepts

Key concepts are important abstractions within the domain that have interesting attributes and behaviors. These abstractions often do not have physical realizations, but must nevertheless be modeled by the system. Within the user interface domain, a window is a key concept. In the banking domain, an account is a key concept. In an autonomous manufacturing robot, a task plan is the set of steps required to implement the desired manufacturing process, a key concept. In the design of a C compiler, functions, data types, and pointers are key concepts. Each of these objects has no physical manifestation. They exist only as abstractions modeled within the appropriate domains as objects or classes.

Identify Transactions

Transactions are objects that represent the interactions of other objects and must persist for a nontrivial period of time. Examples of transactions include bank-account deposits and withdrawals, elevator requests, target designations for fire control systems, and objects that manage reliable message delivery. In many cases, the transactional objects disappear once the interaction has concluded (so-called volatile transactions) while in other cases, the transactional objects must be retained for long periods of time (also known as persistent transactions).

Identify Persistent Information

Persistent information is typically held within passive container objects such as stacks, queues, trees, tables or databases. Configuration data for various devices is one kind of persistent data. Many systems also acquire information from their environments and store this information. Of course, when information must be stored, objects hold that information in their attributes. Examples include a system that remembers patient data, user names and passwords, calibration tables for sensors, or flight data for a black-box recorder.

Identify Visual Elements

Many real-time systems interact directly or indirectly with human users. Real-time system displays may be as simple as a single blinking LED to indicate power status, or as elaborate as a full Windows-like GUI with buttons, windows, scroll bars, icons, and text. Visual elements used to convey information to the user are objects within the user-interface domain. These visual elements are themselves objects, and have both attributes and behavior.

Identify Control Elements

Control elements are entities that control other objects. These are specific types of causal objects. Some objects, called composites, often orchestrate the behaviors of their part objects. These may be simple objects or may be elaborate control systems, such as:

- PID control loops
- Fuzzy-logic inference engines
- Expert-system inference engines
- Neural-network simulators
- State-based systems
- Algorithmic systems

Apply Scenarios

The application of use-case scenarios is another strategy to identify missing objects as well as to test that an object collaboration adequately realizes a use case. Using only known objects, step through the messages to implement the scenario. The object collaboration structure must support the elaborated use-case scenarios, including providing the information to be manipulated and the services realizing the messages in the scenarios. This is one of the most useful strategies because it allows you to identify elements and behavior missing from your collaboration.

In the course of the exercises in this chapter, we will apply each of these strategies to identify the objects within one or more collaborations. You will need to refer to the written problem specification in the appendices as well as the previous requirements models. In an actual project, usually anywhere from two to four of these strategies is sufficient on any given use case.

Problem 5.1   Apply Nouns and Causal Agents Strategies

In this problem we will limit ourselves to applying the "underline the nouns" and "identify causal agents" strategies. The first of these is a very common strategy although, in my experience, the least helpful. The application of this strategy is to simply read the parts of the problem statement that apply to the use case in question, and then underline the nouns and noun phrases. These noun phrases are our candidate objects and classes.[2] The difficulty with the strategy, besides the fact that it only finds the explicitly stated objects, is that it also identifies attributes, actors, objects we don't care about, and synonyms for already identified elements. Thus, the list requires careful pruning and won't be complete in any case. Also be aware that, in this book, we have very short problem statements. In many systems projects, the problem statements will be much longer and complete, possibly running to hundreds of pages for large systems. However, the strategy can be used even in those cases, because we apply the strategy a use case at a time.

The causal agent strategy looks for autonomous and triggered behaviors in the system and tries to identify the sources and targets of those behaviors.

The first part of this problem is to apply these strategies to—you guessed it—the Roadrunner Traffic Light Control System. Because the system is relatively simple, we'll apply the strategy to two system-level use cases: "Detect Vehicle" and "Fixed Cycle Time Mode" (a subclass of the "Manage Traffic" use case). The problem statement relative to the Detect Vehicle use case is shown below:

The Vehicle Detector

Three types of Vehicular Detectors shall be supported: subsurface passive loop inductors (SPLIs), above-surface infrared sensors (ASIs) and above-surface radars (ASRs).

Subsurface detectors shall use a wired interface to communicate with the controller, while ASIs and ASRs shall support both wired and secure wireless communication. All vehicle detectors shall be able to perform vehicle counting.

In addition, ASIs and ASRs shall be able to receive directional transmissions from priority vehicle and emergency vehicle transmitters. The maximum range of such reception shall be no less than 250 feet and no more than 1000 feet.

Figure 5.2 shows the relevant measures for both ASI and ASR detectors. When a vehicle enters the detection area (shown as the shaded area in the figure), the detector shall report the presence of a vehicle. Separate detectors are used for each lane in each direction.

---

[2]   It is true that you can also identify verbs to find behavior and adverbs for quality of service constraints (especially ones with quantitative values), but our initial focus is to identify the objects and classes.

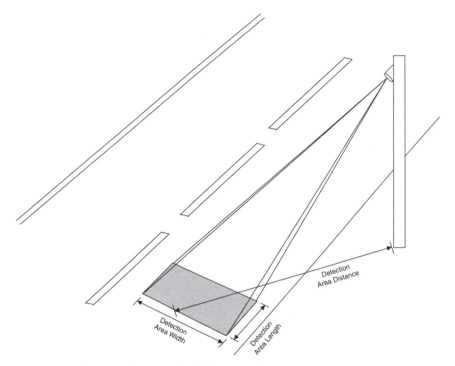

Figure 5.2 Infrared and radar vehicle detector

The part of the problem statement that applies to the fixed cycle mode is shown below:

Mode 2: Fixed Cycle Time

Mode 2 is the most common operational mode. In this mode, the lanes cycle GREEN-YELLOW-RED in opposite sequences with fixed intervals. The system shall ensure that, if any traffic light is non-RED, then all the lights for cross traffic shall be RED and pedestrian lights (if any) shall be set to DON'T WALK. Note that the turn lane times and/or pedestrian times are only valid in this mode if (1) the turn lane and/or pedestrian parameter is set TRUE in the RIC system parameters and (2) if a signal from the appropriate detector determines the existence of waiting traffic for the turn or pedestrian light.

The durations of the light times shall be independently adjustable by setting the appropriate parameters (see below). Note that in the table the values in parentheses are defaults.

Table 5.2 Mode 2 parameters

Parameter	Value range	Description
Reset Parameters	FALSE, TRUE	(FALSE) Sets all the parameters for Mode 2 to defaults
Primary Green Time (PG2)	10 to 180 seconds	(30) Length of time the primary green light is on
Primary Yellow Time (PY2)	2 to 10 seconds	(5) Length of time the primary yellow light is on
Primary Red Delay Time (PR2)	0 to 5 seconds	(0) Length of time between when primary red light is turned on and the secondary green light is activated
Primary Walk Time (PW2)	0 to 60 seconds	(20) Length of time the primary WALK light is on when the primary GREEN light is activated
Primary Warn Time (PA2)	0 to 30 seconds	(10) Length of time the primary FLASHING DON'T WALK light is on after the WALK light has been on
Primary Turn Green Time (PT2)	0 to 90 seconds	(20) Length of time the primary turn light is GREEN. Note: only valid when the Primary Turn Light parameter is TRUE.
Primary Turn Yellow Time (PZ2)	0 to 10 seconds	(5) Length of time the primary turn light is YELLOW. Note: only valid when the Primary Turn Light parameter is TRUE.

The default values depend on the system configuration.

Table 5.3 Default cycle times for Mode 2

Turn Lane	Ped Signal	Green	Yellow	Red	Walk	Don't Walk	Turn Green	Turn Yellow
F	F	30	5	0	0	0	0	0
T	F	50	5	0	0	0	15	5
F	T	50	5	0	15	5	0	0
T	T	50	5	0	15	5	15	5

The values in Table 5.3 are true for each direction, independently. Thus, if the primary road has a car waiting in its turn lane and a pedestrian walking, but the secondary road has neither, then the following timing diagram represents the cycle times for simultaneous turn lane mode (i.e., the turn lanes in both directions for a road turn together and the straight traffic doesn't begin until the turn lanes have cycled to Red).

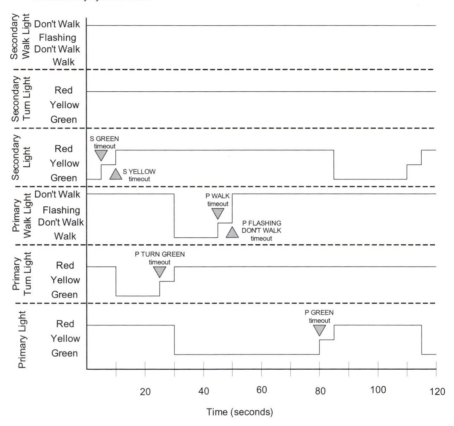

Scenario: Pedestrian and turn lanes enabled 001

Preconditions: Secondary light is GREEN, Primary is RED; Car waiting in primary turn lane; pedestrian waiting for primary walk signal; 5s left in Secondary cycle GREEN

Figure 5.3 Timing diagram for Mode 2 example

The first part of the problem is, then, to apply the Underline the Nouns and Identify Causal Agents strategies to this part of the Roadrunner specification and construct the resulting analysis object model.

The second part of this problem applies the strategies to the CUAV. The CUAV is a much larger system than the traffic light control system—probably two or three orders of magnitude larger (i.e., 100 to 1000 times larger). In such cases, it is common to apply the object identification strategies at the subsystem level rather than the system level. This fits well into the normal work allocation for systems as well. It is common to hand off subsystems to different (interdisciplinary) subsystem teams with a specification of what that subsystem must do. We recommend that the subsystem specification include the subsystem use cases, the system architecture into which the subsystem must fit, and a complete specification of the interfaces the subsystem must provide or may require from other elements (subsystems or actors). These subsystem-level use cases are much more tractable in scope.

We will apply the strategies to a single subsystem level use case—Reconnaissance Management. The use cases for this subsystem are shown in Figure 5.4. For this exercise we will use the "Acquire Image" use case. As an aside, note that the peer subsystems in the figure are prefaced with a lower-case "a"—this is used to indicate that we are using the subsystem as an actor from this point of view. In reality, it is a different metaclass with a very similar name. This allows us to provide behavior for the actors (called "instrumenting the actors") for simulation and test without modifying the actual subsystems they represent.

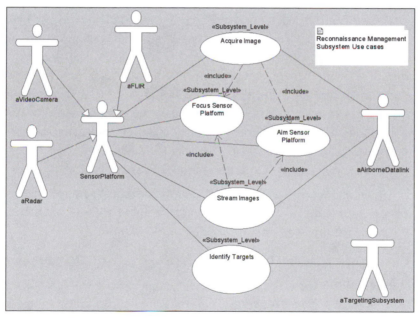

Figure 5.4 Reconnaissance Management use cases

It should be noted that one of the difficulties of applying the Underline Nouns strategy in big systems is that the textual specification may not be organized in a fashion to facilitate a per-use-case application of the strategy. It may require reorganization of the textual specification or multiple passes through the document as a whole to identify all the sections related to the use case under consideration. If the document is organized by subsystems and then by use, this problem may disappear but, in my experience, that is an unusual organization.

The Unmanned Air Vehicle (UAV)

The Coyote UAV is meant to be a multipurpose reusable UAV with multi-mission capability. It is meant to operate at an altitude of up to 30,000 feet with ground speeds of up to 100 knots (cruise) and 150 knots (dash) and carry a payload of up to 450 lbs for durations in excess of 24 hours. The Coyote is meant to fly unimpeded in low-visibility environments while carrying either reconnaissance or attack payloads. While controllable from the ground station CMPCS, it is also capable of flying complex flight plans with specific operational goals of systematic area search, ground route (road-based) search, and orbit surveillance of point targets. Coupled with manned control from the ground, the Coyote provides sustained 24-hour flight with real-time visual, infrared or radar telemetry, with target recognition preprocessing. Communications are jam-resistant, although need not be antijamming in a high ECM environment. Control commands shall be encrypted while telemetry data can be compressed but unprotected. Telemetry rates for visual telemetry support 30 frames-per-second (fps) at 640 x 400 resolution. Range of flight is meant to be fully supported within line of sight (LOS) range but since the Coyote also has the ability to be passed among different CMPCSs, its range is considerably greater than LOS. For navigation, the Coyote has on-board Global Positioning System (GPS) based-navigation as well as being directly controllable from the ground station.

Unlike many smaller UAVs, the Coyote does not require specialized launch and recovery vehicles. It can use a short runway for either automated or remote-controlled takeoff and landing.

The Coyote Mission Planning and Control System (CMPCS)

Mobile CMPCS with capability to control up to four UAVs with a manned control station per UAV that fits into a smaller towable trailer. Each control station consists of two manned substations, one for controlling the CUAV and one for monitoring and controlling payloads. If desired, both functions can be slaved together into a single control substation. Control of the aircraft shall consist of transferring navigational commands which may be simple (set altitude, speed, direction), operational (fly to coordinate set, orbit point, execute search pattern, etc.), planned (upload multisegment flight plan) or remote controlled with a joystick interface. Stable flight mechanics shall be managed by the aircraft itself but this can be disabled for remotely controlled flight.

The CMPCS displays real-time reconnaissance data as well as maintaining continuous recording and replay capability for up to 96 hours of operation for four separate CUAVs. In addition, with attack payloads, the Coyote can carry up to four Hellfire missiles with fire-and-forget navigation systems.

The Unmanned Air Vehicle (UAV)

...

Mission Modes

Beyond flight modes, CUAV shall be designed for highly flexible mission parameters. Normal mission modes include:

- Preplanned reconnaissance
- Remote-controlled reconnaissance
- Area search
- Route search
- Orbit point target
- Attack

A mission can consist of any number of sequential submissions, each operating in a different mission mode, depending on the current payload.

The Coyote Mission Planning and Control System (CMPCS)

The CMPCS is housed in a 30 x 8 x 8 triple-axis trailer that contains stations for pilot and payload operations, mission planning, data exploitation, communications, and SAR viewing. The CMPCS connects to multiple directional antennae for communication with the CUAVs. All mission data is recorded at the CMPCS since the CUAV has no on-board recording capability. The CMPCS has a UPS that can operate at full load for up to four hours in addition to using commercial power or power generators.

A single CMPCS can control up to four CUAVs in flight with one station per CUAV. Each CUAV control station provides both pilot and payload operations with separate control substations, although both functions can be slaved to a single substation for low-vigilance use.

For the reconnaissance payloads, the CMPCS shall provide enhanced automated target recognition (ATR) capability for all surveillance types—optical, infrared, and radar. While the CUAV has a rudimentary capability, the CMPCS provides much more complete support for the quick identification of high-value targets in the battlefield. This capability is specifically designed to identify mobile and time-limited targets that may only be exposed for brief periods of time before they go back into hiding. The system is expected to provide high clutter rejection and a low false-positive error rate. The ATR shall be able to identify and track up to 20 targets within the surveillance area, with likely identification and probability assessments for each. In addition to the ATR, the payload operator can add targets visually identified from reconnaissance data or gathered from other sources. The battlefield view can be transmitted over links to remote command staff for tactical and strategic assessment.

Image Acquisition and Processing

Images may be acquired from all three sensor platforms—optical, FLIR (forward-looking infrared), and SAR (synthetic-aperture radar). Optical and FLIR are passive systems using emitted energy from terrain or targets to gather information. The SAR is an active sensor, painting relatively stationary targets with energy (in the microwave range) and using the reflection of these pulses to determine reflectivity and altitude. The optical and FLIR resolution for single images may be as high as 1900 x 1600 resolution while real-time video for all sensor platforms is limited to 640 x 480 resolution at a rate of 30 fps. Streaming imagery shall be sent with enough redundancy so that complete loss of random frames shall not affect the quality of other frames. The sensor platforms may be focused at any range from 10 meters to infinite and may be zoomed up to 100x actual. The sensor platforms shall be mounted on a gimbaled assembly so that the system can be aimed without affecting the attitude of the CUAV. The FLIR includes a laser range finder to determine target range so that it can be used in fire control applications. The SAR shall emit a series of pulses meant to emulate the behavior of much larger physical aperture antenna; the images from the SAR are a combination of timed radar surface reflections to be combined into an SAR image in the ground station using Fourier transforms. Thus, a single SAR image results from a set of images each resulting from a single radar pulse from the SAR platform, but combined in the ground station. Aiming the SAR is done through the use of Doppler sharpening, limiting the amount of information that must be transmitted to the ground station to construct the SAR image. The use of two pulse emitters in the SAR platforms allows the interference patterns to be constructed, providing altitude determination as well as radar reflectivity data.

Images may be compressed using lossy or nonlossy methods to minimize communication bandwidth requirements. The JPEG 2000 compression standard shall be used; for streaming video the associated MJP2 standard shall be used. The compression may be set dynamically by the payload operator to be 0% to 80% with the default setting to be nonlossy 50% compression. The imaging system is required to achieve the desired compression only within 20% of the requested due to the variances in the image contents. The selection of lossy or nonlossy compression shall be determined automatically by the imaging system, switching to lossy compression only when the desired compression rate cannot be achieved using lossless compression.

For this part of the first problem, apply the Underline the Nouns and Identify Causal Agents strategies to the "Acquire Image" use case, and draw the resulting object analysis model.

Problem 5.2 Apply Services and Messages Strategies

The Services strategy looks for the services or "operational contracts" that the system containing the collaboration provides. Each of these services must either be met by a single object in the collaboration or be decomposed into nested services (possibly multiple times), each of which must be provided by a single object. The Messages strategy applies the same principle to messages and information flows; every message or information flow must be either provided or received by an object in the collaboration. In some sense, these strategies are the same. However, in the former case, the strategy is applied when the system containing the collaboration is specified in terms of a set of operational contracts (services) and, in the latter case, the strategy is applied when the system is specified by information flows.

For this problem, apply these two strategies to the same use cases as before for the Roadrunner and Coyote systems. Because this level of detail is missing in the work that precedes the exercise, both operational contract and information flow views of the Roadrunner Traffic Light Control System and CUAV systems are given below.

The two use cases under consideration for the Roadrunner system are "Detect Vehicle" and "Fixed Cycle Time Mode."

The services for the subsystem architecture of the Roadrunner Traffic Light Control System are shown in Figure 5.5. We could explicitly show the interfaces attached to the ports if desired, but chose not to in this case.

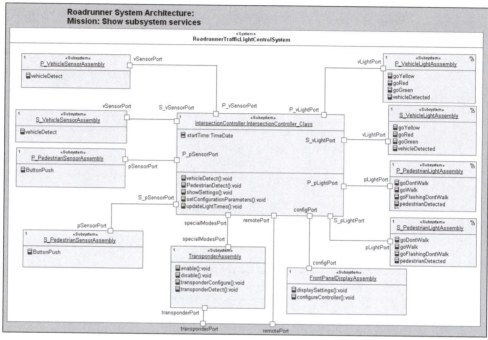

Figure 5.5 Roadrunner services

The next figure (Figure 5.6) shows the same structure but focuses on the information flow.

For the UAV, consider the Acquire Image use case and apply this strategy to the identification of objects in the Reconnaissance Management subsystem. There are a number of services associated with acquiring an image, and certainly data must be passed. Draw the class diagram derived from applying this strategy.

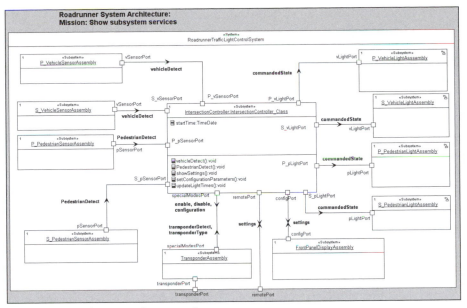

Figure 5.6 Roadrunner information flows

Problem 5.3 Apply Real-World Items and Physical Devices Strategies

These two strategies are related in the sense that they seek to identify objects and classes that represent things that exist in the physical world. The first of these—real-world items—seeks to identify objects in the real world that have information that must be managed in the system or are resources that must be managed. For example, in a banking system, a "customer" is clearly an object in the real world but we must maintain information about that customer, including Name, Address, and Tax ID. If we think about a turreted 20mm machine gun aboard, say, a AH-1 Super Cobra attack helicopter, the rounds of ammunition are a resource that must be tracked and managed and its targets are certainly real-world elements that must be represented and managed.

The latter strategy identifies physical devices. These devices are not devices external to the system but, rather, a part of it. Some beginning modelers will model all physical devices that the system must monitor and control as actors, but that is not always the best approach. Using the "Rule of the Box," a piece of hardware is only an actor if it is not integrated with the software into a shipped system.[3]

---

[3]   The "Rule of the Box," you will no doubt remember, states that if the hardware is "in the box shipped to the customer" then it is not, by definition, an actor. If it is provided by the customer and connected to the system you ship, then it is likely to be an actor.

Apply these strategies to the same use cases as the previous problem.

Problem 5.4　Apply Key Concepts and Transaction Strategies

The Key Concepts strategy is the antipode of the real-world items strategy. It seeks to find the essential concepts of a domain of discourse, particularly when these elements are abstractions and have no physical manifestation. An "account" in a banking system, or "thread" in an operating system are both examples of this strategy.

A transaction is the reification of an interaction (among objects) into an object itself. This is done when that interaction has a lifetime of its own and persists at least until the transaction is completed. A "withdrawal" from a bank account is a transactional object because it must be remembered for a period of time and reported to the customer for account reconciliation. A request for an elevator to go to a particular floor is also a transactional object because it must be remembered until the elevator actually arrives at the floor.

For this problem, apply these two strategies to the Roadrunner Traffic Light Control System Detect Vehicle use case and also to the Coyote UAV Acquire Image use case.

Problem 5.5　Apply Identify Visual Elements and Scenarios Strategies

In this problem we will consider the Identify Visual Elements strategy along with the Scenarios strategy. In systems with nontrivial user interfaces, these strategies often work well together, because one of the issues that arises when considering scenarios that involve the human users of the system is how the user interfaces gather and provide information to the internal parts of the system that must deal with and respond to that information.

While all the strategies have their uses, and every analysis problem requires the application of at least two to three strategies, the Scenario strategy is my favorite. It has proven to be, for me at least, the easiest way to construct a verifiably working collaboration of classes to realize a use case. The strategy is actually very simple. Simply start with a use-case scenario. The use case will have a (possibly large) number of scenarios as exemplars, illustrating examples of the use case unfolding as specific messages or events come in to the system. Since these scenarios are "black box," there is only a single lifeline representing the system (or use case). We will simply elaborate object roles to show how the scenario unfolds at the object, rather than the system, level. In small-scale systems, such as the Roadrunner Traffic Light Control system, system-level use cases are used. In large-scale systems, such as the Coyote UAV, this is normally done at the subsystem level of abstraction.

This elaboration can be done "in-line" or by decomposition. By "in-line," I mean that I copy the original scenario and start adding object lifelines to the copy. The decomposition approach is done by decomposing the lifeline into a more detailed scenario, a feature added in UML 2.0 and supported by Rhapsody.

During the object-level scenario elaboration, I will uncover objects, services, and parameters (data). When I discover that I need to have a service performed somewhere inside the system, I ask myself the following questions:

- What object has the information necessary to perform this service?
- What object has the proper interfaces necessary to perform this service?
- What object has the responsibility to perform this service?

I then add the object, type it with a class, add the service to the class, and create an association between the client and the server classes.

By this means, I add objects, services, and relations to detail this scenario. I may start at the beginning of the scenario, or somewhere in the middle—perhaps at a very important part of the use case or some part that I feel I have a good handle on. As I add objects to the scenario, I in tandem add relevant classes to a class diagram, filling in the operation calls and event receptions to handle the message, attributes to supply the information required for the processing, and associations necessary to support the sending of messages within the collaboration.

Figures 5.7 and Figure 5.8 illustrate how this elaboration takes place. Figure 5.7 shows the high-level interaction. The lifeline in the middle is the use case, but it can just as easily be the "System" or another high-level object. The point is that this element internally contains parts (typed by classes) that interact to provide the high-level behavior shown.

In the elaborated copy of this sequence diagram (Figure 5.8), detailed elements "inside" the system of interest are shown with their detailed interactions. Note that we retain the original high-level use case or system lifeline. This isn't strictly necessary—we could "reroute" the messages to the internal parts—but the advantage for the former approach is that it becomes trivial to map this interaction back to the

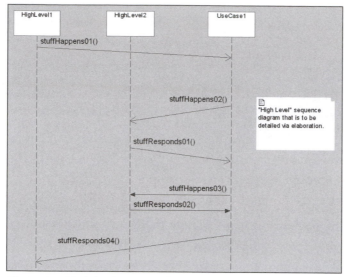

Figure 5.7: High-level sequence diagram before elaboration

original, thus demonstrating the equivalence of the two. If we reroute messages in the latter diagram we can still demonstrate equivalence, but it becomes less trivial and therefore we are more likely to make errors. In any event, the point of the latter diagram is to show how the low-level internal parts collaborate to achieve exactly the behavior described in the high-level sequence diagram.

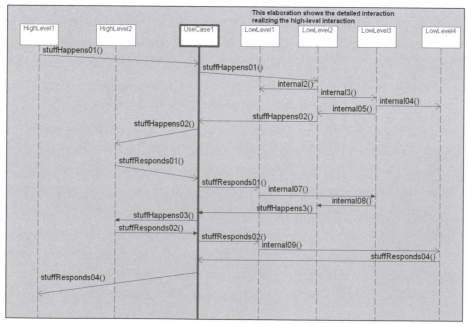

Figure 5.8 High-level sequence with elaborated detail

For the first part of this exercise, continue on with the Roadrunner Traffic Light Control System "Detect Vehicle" use case. Copy the original sequence diagram(s) and add the object lifelines as you do so—the "scenario elaboration" approach. Provide both the elaborated sequence diagram(s) and the class diagram when you're done.

At first glance, the Detect Vehicle use case doesn't seem to be an ideal candidate for this strategy. After all, the "scenario" consists of a single message "vehicleDetect" from the vehicle to the system (well, almost anyway). However, a use-case scenario should never be a single message. A scenario with a single message should always be modeled as an element of a large scenario. So, I decided to include the configuration of the detectors as a part of the scenarios. Each of the detectors must be configured differently. As a starting point for this exercise, use the following three scenarios: Figure 5.9, Figure 5.10, and Figure 5.11 for the starting point of the elaboration. You'll no doubt notice the essential similarity between the scenarios, but each detector type has its own configuration parameters. Additionally, the lanes can be configured separately or all at once. For example, in Figure 5.9, which represents the scenario for the passive loop inductor, the first parameter is the laneID. This parameter indicates for which lane you're configuring the detector. Additionally, the sensitivity for the detector is set by the second parameter. Since passive loop inductors look for a change in resistance, this sets the trigger point for the detection to one of 11 values (0 essentially turning off the detector). There is a special "sentinel value" for the laneID that configures or enables all the detectors at once, if desired.

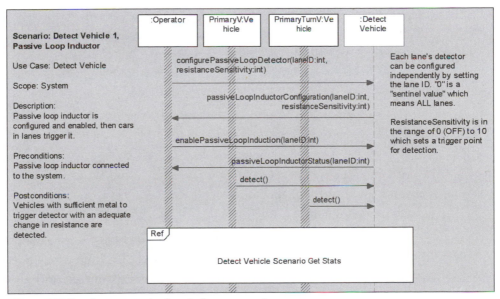

Figure 5.9 Roadrunner passive loop inductor scenario

What happens if the laneID or the sensitivity provided is out of range? In that case (scenario not shown—another recommended "exercise for the reader"), the invalid command is discarded and the previous settings are returned to the operator. As is normal, there are more fault or "rainy-day" scenarios than normal or "sunny-day" ones. In the actual application of the strategy on a real project, you would elaborate every scenario, especially in the sections of the scenario that are unique.

In Figure 5.10, the infrared detector requires not only the laneID and sensitivity, but also the low and high frequency, since this detector can be set to look for changes within a frequency range. The low- and high-frequency ranges are selected from 11 predefined values, with the constraint that the high frequency must be greater than or equal to the low-frequency setting. What happens if the operator violates this constraint? Hmmm, sounds like another scenario!

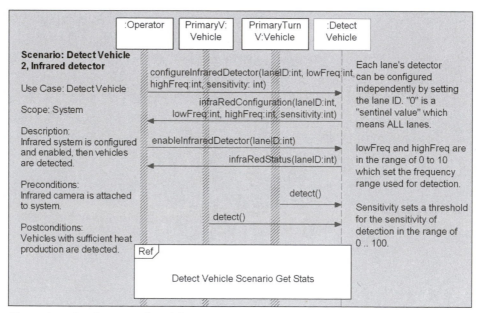

Figure 5.10 Roadrunner infrared detector scenario

The radar detector is the only active emitter. It only emits a specific frequency but both the strength of the emitter and the sensitivity of the detector can be set, as illustrated in Figure 5.11.

Each of these scenarios references another—one that returns statistics to the user, including average vehicle count per hour and total count since reset. This latter scenario is referenced by all three scenarios because the generation of statistics

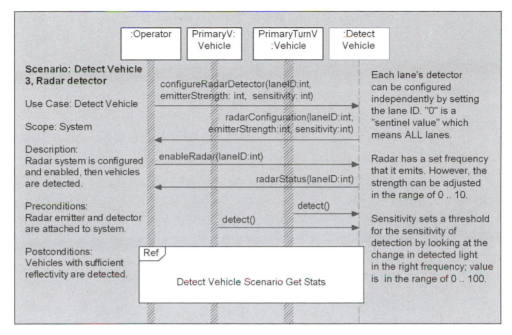

Figure 5.11 Roadrunner laser detector scenario

depends only upon the detection of a vehicle and not how they are detected. This shared scenario is shown in Figure 5.12.

The astute reader will note that the inclusion of the formal parameter flowID, along with the diagrammatic comment describing the effects of the different values

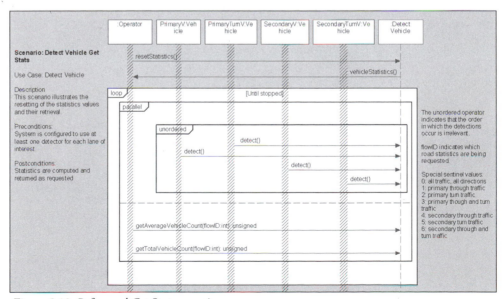

Figure 5.12 Referenced Get Stats scenario

of this parameter, is just a sneaky way of putting many different physical scenarios on a single sequence diagram, as is the use of the unordered interaction operator. The unordered operator simply states the messages within the operator can occur in any order whatsoever—that is, there is no causality in the order in which they arrive. I could have, in principle, simply created a different sequence diagram for each order of arrival of detections from four different lanes and combination of flowIDs, resulting in many highly similar but slightly different scenarios. I find the use of these operators a more parsimonious means of achieving the same end.

As far as the human interface goes, the external UI has already been planned out and provided in the requirements specification. The details can be seen in Appendix A, but the organization of the UI is shown in Figure 5.13.

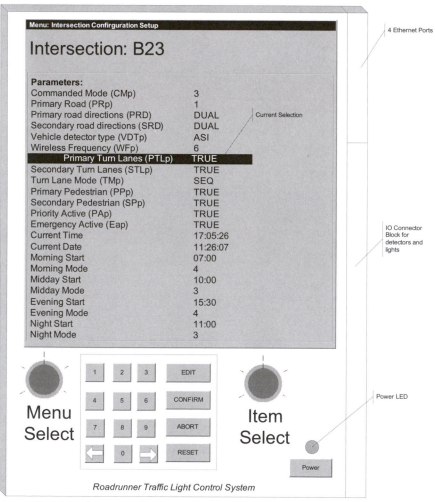

Figure 5.13 Roadrunner traffic light control system front panel display

This front panel display must support the selection of the appropriate parameters for this use case (notably the configuration parameters for the sensors) and for the display of those same parameters and the computed statistics.

For the latter part of this exercise, let's take the scenarios of the Acquire Image use case for the CUAV. In this case, though, add a reference sequence diagram to represent the use-case lifeline at a more detailed level. For consistency, be sure that for every message on the use-case sequence diagram going into the use-case or system lifeline, there is a corresponding message of the same name leaving the Environment lifeline in the more detailed scenario and for every message leaving the use-case or system lifeline, there is a corresponding message entering the Environment lifeline in the more detailed use case.

This decomposition is easy to do in Rhapsody, Simply double click on the lifeline (or right click and select Features) to open the Features dialog. The dialog box contains a Decomposed field, which is the reference to the nested sequence diagram. From that dialog, you can either link to an existing sequence diagram or create a new, empty one.

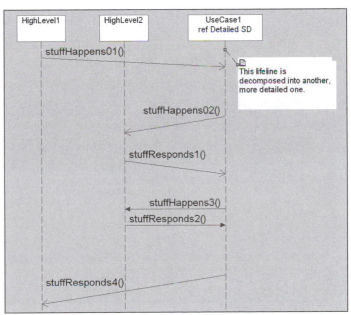

Figure 5.14 High-level sequence diagram with decomposed lifeline

The "UseCase1" lifeline in Figure 5.14 is decomposed into the more detailed interaction in Figure 5.15. The key to the consistency between the high-level and decomposed sequence diagrams is the use of the ENV (environment) lifeline in

Figure 5.15. It provides the "glue" between the levels of abstraction in the two diagrams so be sure that the messages entering and leaving the UseCase1 lifeline match exactly the messages entering and leaving the ENV lifeline on the decomposed sequence diagram.[4]

Decomposing sequence diagrams in this fashion was added in UML 2.0 and is now the preferred way to add lower-level detail of interactions in your models. This is especially true for large systems in which you might have to decompose high-level sequence diagrams through potentially multiple levels of abstraction (e.g., system level → subsystem level → component level → collaboration level).

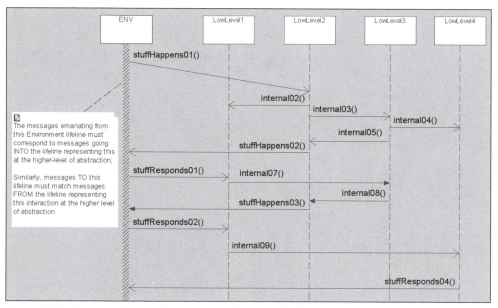

Figure 5.15 Sequence diagram for decomposed high-level lifeline

For the Reconnaissance Management subsystem Acquire Image use case, use the use-case scenario shown in Figure 5.16 to start your object identification. In your solution, provide both the decomposed sequence diagram and the resulting class diagram.

---

4   And yes, you have to maintain this "glue" manually. If you add messages in one diagram that imply changes in the other, you have to do that yourself.

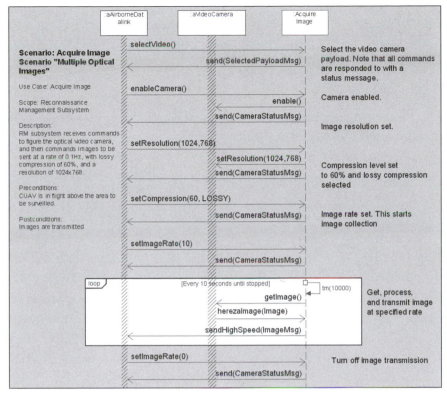

Figure 5.16 Acquire Image scenario

Problem 5.6 Merge Models from the Various Strategies

In this last problem of the chapter, you must merge together the different models that have arisen from the application of the different strategies. In actual use, the models would be most likely incrementally constructed by taking what's already there and adding objects, classes, relations, and features identified in additional strategies. In this pedagogical chapter, we have applied each strategy more-or-less independently. Therefore, for this problem, merge together the solutions from the different strategies and resolve any incompatibilities arising from different approaches to represent the same concept for both the Roadrunner Traffic Light Control System and for the Coyote UAV. The resulting use-case collaborations should be shown on a single class (or structure) diagram for each system.[5]

---

[5]   By way of a workflow hint, create a new diagram and drag or copy elements from other diagrams to it, then use the Complete Relations Tool (menu Layout → Complete Relations) to draw the existing relations among the elements, then manually adjust the layout to get a pleasing graphical depiction.

For the Roadrunner Traffic Light Control system, we've worked, in part, on two different use cases, Fixed Cycle Time Mode and Detect Vehicle. The system includes the elements that collaborate to realize both these use cases. These use-case collaborations have been shown as two independent class diagrams and they will continue to be shown in that way. Why?

The reason is that to scale the use of modeling techniques to real-world problems, we must have some criteria for deciding what goes on which diagrams. The criteria I've been (strongly) recommending is the notion of a "diagram mission"—i.e., each diagram should have a single important concept that it's trying to show. If we maintain the diagrams as independent, then each maintains the purity of its purpose. However, there almost always is overlap between different collaborations in terms of the elements they employ. That's perfectly fine, provided that the overlapping elements are the same elements in all relevant views. A tool with an underlying model repository—such as Rhapsody—will maintain the integrity of the classes, objects, operations, etc. used in multiple views. So for this exercise, show the Fixed Cycle model class diagram separately from the merged Detect Vehicle use case. Identify to yourself any elements they have in common and ensure that these shared elements meet both purposes for which they are intended. Note, however, that this chapter's focus on identifying objects and classes and class structure for Fixed Cycle model is pretty trivial—the interesting part is in the behavior of these elements, as captured with the state machines for those class elements (PedestrianLightAssembly and VehicleLightAssembly, primarily).

The CUAV is a larger scale system, so we've narrowed our focus onto a single use case (Acquire Image) for a specific subsystem (Reconnaissance Management). One can imagine teams of people working independently, possibly in different areas of the world, to develop the use cases for their own subsystems. As long as they meet the interface, functional, and quality of service requirements, those subsystems will plug into the system architecture seamlessly. This is best done iteratively, bringing the partially developed subsystems together repetitively at well-defined integration points in an iterative development cycle, and formally testing the system functionality as it grows over time.

For this problem, though, we'll limit ourselves to merging together the collaborations derived from applying the different strategies to a single use case.

Finally, for both sample problems, answer the following questions:

- Which strategy worked best for you, and why?
- Which strategy worked the least well for you, and why?

- Did some strategies seem to be better identifying elements in one problem or the other? Can you generalize that so that you know when to apply which strategies?
- What combination of strategies do you think will be most effective for you?

Looking Ahead

We have used a number of different strategies to identify the "essential" or "analysis" objects and classes. These strategies are only partially orthogonal; that is, they identify some, but not all, of the same model elements. In practice, the application of two or more strategies is required to identify all of the objects in the analysis model. These objects are "essential" in the sense that their presence is required by the nature of the problem being solved. A microwave oven had better have a microwave emitter, a track manager requires the notion of a track, and a navigation system needs the concepts of position, flight plan, and waypoint to do its job.

The analysis model exactly corresponds to the idea of a platform independent model (PIM) in the OMG's model-driven architecture (MDA).[6] It is devoid of design decisions. The analysis model is driven by the functionality requirements specified by the use cases of the system. The objects, classes, attributes, operations, and relations identified may be optimized in many different ways using many different design approaches with many different technology decisions to construct the platform specific model (PSM).

While analysis is all about identifying elements required to meet the functional requirements, design is all about optimizing the analysis model to meet performance requirements and other design optimization criteria. Different optimization goals or different technology selections result in different PSMs from the same PIM. The advantage of this approach is independence of your analysis models from "technology churn" and the ability to adapt the analysis model to meet different optimization criteria, to use different technologies, and to execute on different platforms. In this way, the approach insulates you from technology churn and increases the interoperability of your system and its portability to new platforms. For systems that have a long life—such as military and aerospace systems—or systems that are part of product families and so have high reuse requirements, the MDA approach is both highly practical and effective. How to achieve this optimization through design is the subject of the next two chapters.

---

[6] See *http://www.omg.org/mda/*.

6

Architectural Design

Overview

What is architecture? IEEE defines it as follows:[1]

> *"An architecture is the fundamental organization of a system embodied in its components, their relationships to each other, and to the environment, and the principles guiding its design and evolution."*

So what does that mean in practice? In the Harmony process we define architecture to be the set of strategic design decisions that specify how the elements in the system are organized and interact. The key terms in our definition are strategic and design. In the Harmony process, design is all about optimization. The analysis model is driven primarily by the functional requirements of the system—what the system needs to do to be correct. Design is driven by the quality of service

---

[1]   IEEE STD 1471-2000.

141

requirements—how well those functions must be achieved—and other optimality characteristics collectively known as "design criteria."

An analysis model may be optimized in almost infinitely different ways to achieve different optimization goals. For example, memory usage can be optimized at the expense of worst-case performance, or reusability can be optimized at the expense of development time. The analysis model specifies what must be present for the solution to be correct; design specifies a solution that is optimal against the criticality-weighted set of design criteria. A set of collaborating objects identified in object analysis can be run in a single thread or it can be run with one thread per object. The number of threads, their properties, and which objects execute within those threads is determined by optimizing the execution of those objects against the weighted set of design criteria.

The other key term in the Harmony definition was *strategic*. By strategic, we mean that all system elements must be aligned with the architectural decisions. Since architectural design decisions are an attempt to optimize the system at a gross, or overall, level, such decisions are strategic. The Harmony process applies design at two other levels of abstraction as well. Mechanistic design is focused on the optimization of collaborating sets of objects—specifically, at the level of use-case collaborations. Those "mechanistic" design decisions are local only to those collaborations specifically addressed, and so their scope is an order of magnitude—or more—smaller than the strategic architectural design decisions. The smallest scope for design in the Harmony process is *detailed*. Detailed design focuses on the optimization of individual classes and objects. Mechanistic and detailed design are the subject of the next chapter.

Regardless of the scope of the design effort, design proceeds largely through the application of design patterns.[2] Design patterns are not magic bullets by any means; they are simply the codification of what good designers already do. Design patterns capture generalized solutions to problems that reoccur in a variety of application contexts. The very best designers already reapply previously proven design solutions, even if they proceed in an intuitive manner. The design pattern approach captures those solutions in reusable ways that facilitate their reapplication to other problems.

---

[2]   For more detail on this topic, the interested reader is referred to the author's book *Real-Time Design Patterns: Robust Scalable Architectures for Real-Time Systems*, Addison-Wesley, 2002.

The Harmony design approach is very design-pattern oriented. A design pattern has four very important properties:

1. The problem context
2. What the pattern seeks to optimize
3. The solution (the pattern itself)
4. The pattern's consequences

The problem context identifies what are the characteristics of the problem required for the pattern to be applicable. These characteristics specify a set of parameters by which you can select suitable problems for the pattern. The optimization criteria are the goals of the pattern—that is, what aspects of the problem this pattern optimizes. The solution is the pattern itself, usually represented as a class or structure diagram, possibly accompanied by statecharts for some of the elements of the pattern, and sequence diagrams show how the elements in the pattern interact to achieve the pattern's optimization goals. Lastly, the consequences are the explicit statement of the pros and cons of applying the pattern. Since the design—and the pattern—is about optimization, it inherently optimizes some aspects at the expense of de-optimizing others. Knowing which aspects are optimized and de-optimized allows for the selection of appropriate patterns for the circumstances.

To apply design patterns in the Harmony process, you typically start with a problem context—a set of structural elements with inherent relations identified in the problem statement or from the problem semantics, such as an analysis collaboration. Then the design criteria are explicitly stated and ranked in order of criticality. Design patterns that optimize the most important of those criteria at the expense of the least important are selected. Lastly, the design patterns are applied by adding in the structural elements (classes and object roles) from the pattern into the original problem context, and adding and adjusting the relations as appropriate for the pattern. This usually results in a certain amount of restructuring of the original elements, a process known as "refactoring."

For architectural design, this process is laid out in Figure 6.1. This is an activity diagram that shows the developer workflow during architectural design. The Harmony process identifies five key architectural aspects:[3] subsystem and component architecture, concurrency and resource management architecture, distribution architecture, safety and reliability architecture, and deployment architecture. These

---

[3] In some systems, there is a sixth aspect—security architecture—as well. It is less common but is reflected in the figure.

aspects are explicitly identified because the computer science literature normally treats these as independent topics and each has its own independent set of design patterns. The overall architecture is the merged sum of the selected patterns from each of these key architectural aspects.

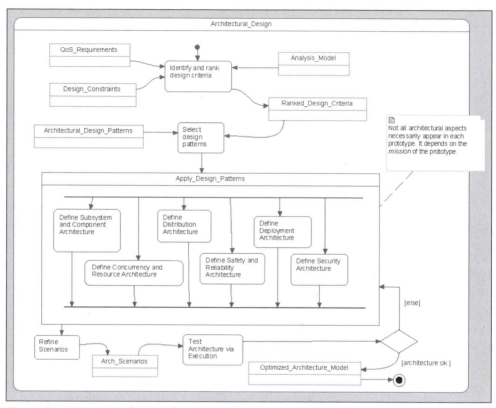

Figure 6.1 Harmony architectural process

So, how does one actually apply a design pattern? The process is best illustrated with a simple mechanistic example. Figure 6.2 shows a collaboration of classes that collaborate together to gather optical data used to determine train speed and then update the data's clients—the text view on the display and for the closed-loop speed controller.[4] This is an analysis-level collaboration and, in fact, executes. It is not necessarily, though, optimal. As identified in the comment in the figure, how do we efficiently update these clients for train speed with new data?

---

[4]   To be clear, a *server* is an object that provides something, either data or services, by request. A *client* is the requestor of those data or services. In the context of Figure 6.2, TrainSpeed is a server for the information needed by clients TextView and SpeedController.

There are several possible solutions. One is to simply have the clients ask for data periodically. This may result in the clients spending many CPU cycles asking for data even though it hasn't changed. Alternatively, the Train Speed class can send the updated values to the clients when the values change. The downside of this approach is that the Train Speed must know, at design time, all of its clients. This solution isn't flexible if we want to dynamically add or remove clients, or if clients might be relocated onto other processors.

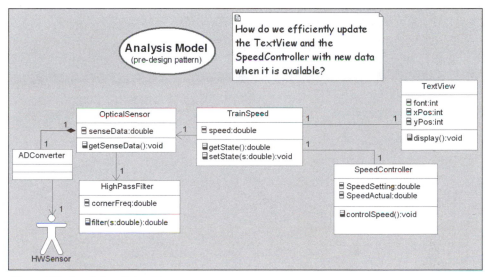

Figure 6.2 Analysis collaboration

What are the design criteria we wish to optimize? Let's say that our two most important design criteria are 1) that the update should be efficient (that is, the clients are only updated when there is new data), and 2) that we wish to be able to easily add clients downstream, possibly even at run-time, without affecting the TrainSpeed and related classes. If we search the literature, we see that the Observer pattern optimizes these two characteristics. Figure 6.3 shows the structure of this pattern.

The Observer pattern works like this:[5] The AbstractSubject has information that might be of interest to some clients, but it doesn't know who those clients are. The clients know where the server is and can connect to it when needed. So far, this is nothing more than the classic Client-Server model. What is special about the Observer pattern is that the server knows how to maintain a list of clients and provides a subscription service so that interested clients can sign up when they want data and

---

[5]   Ibid.

disconnect when they are no longer interested. When the client subscribes, they are typically immediately sent the data, and subsequently sent updates when some update criterion is met. The most common update criterion is when the data changes, but other update criteria might be used, such as sending the data periodically.

As with all patterns, the Observer pattern has two kinds of classes in it. The first are "provided classes." These are classes that come along with the pattern that serve as glue to hold the pattern together and facilitate its execution. In Figure 6.3, NotificationHandle, AbstractSubject, and AbstractObserver are the "glue classes." The second kind are the formal parameters of the pattern, places where you hook in your classes from your analysis models. The ConcreteSubject and ConcreteObserver represent the classes in your analysis model. Thus, to apply the pattern we will add the glue classes and replace the formal parameter classes with the actual classes from the analysis model. The result of this design pattern application is shown in Figure 6.4.

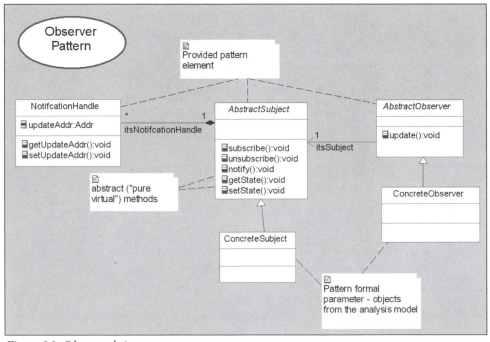

Figure 6.3 Observer design pattern

Figure 6.4 has the elements from the original collaboration, the glue classes, and the classes that replace the formal parameters of the pattern. Generalization is a primary, although not the only, way in which design patterns are applied to the collaboration. Also note that the original collaboration has been refactored slightly;

specifically, the original relations between the TrainSpeed, TextView, and SpeedController class are removed and essentially replaced by a one-way relation between the AbstractSubject and the AbstractObserver.

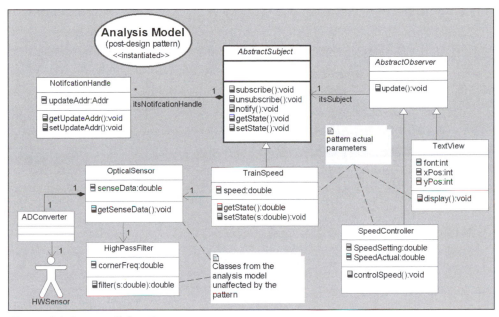

Figure 6.4 Design collaboration

It is not the purpose of this book to replicate all the information in other books on patterns.[6] It is in the province of this book to give you experience in their application, however. To this end, we will take the work we have already done, identify some architectural design criteria, and apply design patterns to it to construct an architectural design model. We will focus here on only three of the architectural views—concurrency/resource, distribution, and safety/reliability.

Problem 6.1   Concurrency and Resource Architecture

The focus of the concurrency and resource management architecture is:

- Identify tasks

- Identify resources

- Map semantic objects into task threads

---

[6]   It is important to remember that this is meant to be a workbook used in conjunction with other books, such as the author's *Real-Time UML, Third Edition*. Thus, we will select patterns to use in the exercises but not fully describe the formidable array of patterns that could have been applied.

- Identify synchronization points
- Specify synchronization algorithms
- Specify how resources will be shared among threads

In the UML, «active» objects are used to model concurrency units.[7] An «active» object is a structured composite object that owns its own thread. It may contain internal parts (object roles typed by classes) that execute nominally in the context of the «active» object's thread. The UML itself doesn't distinguish between heavyweight and lightweight threads since they differ in implementation detail but not really in fundamental semantics. The Harmony process identifies the concurrency architecture using the workflow in Figure 6.5.

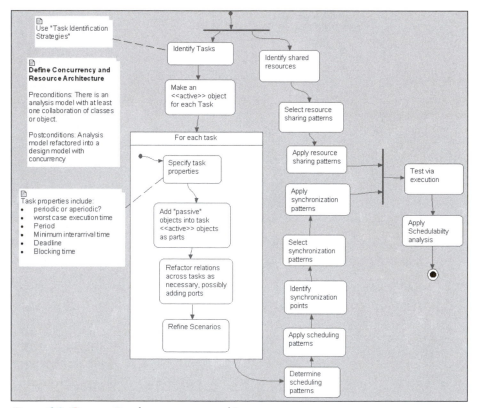

Figure 6.5 Constructing the concurrency architecture

---

[7]   The notation for an «active» class is a class drawn with a heavy border (UML 1.x) or with a double-line border on its left and right edges (UML 2.0).

The four major activities done in the specification of the concurrency and resource architecture are 1) specify the tasks, 2) specify the resources and how they are to be shared, 3) specify the scheduling patterns to be used, and 4) specify how the tasks will synchronize when necessary.

The first job is to find a good set of tasks. In the Harmony process, we do this by using one or more task-identification strategies.

Table 6.1 Task-identification strategies

Strategy	Description	Pros	Cons
Single event groups	Use a single event type per task	Very simple threading model	Doesn't scale to many events well; suboptimal performance
Interrupt Handler	Use a single event type to raise an interrupt	Simple to implement for handling a small set of incoming event types; highly efficient in terms of handling urgent events quickly	Doesn't scale well to many events; interrupt handlers typically must be very short and atomic; possible to drop events; difficult to share data with other tasks
Event Source	Group all events from a single source so as to be handled by one task	Simple threading model	Doesn't scale to many event sources well; suboptimal performance
Related Information	Group all processing of information (usually around a resource) together in a task	Useful for "background" tasks that perform low-urgency processing	Same as Event Source
Independent Processing	Identify different sequences of actions as threads when there is no dependency between the difference sequences	Leads to simple tasking model	May result in too many or too few threads for optimality; doesn't address sharing of resources or task synchronization
Interface Device	A kind of event source	Useful for handling device (e.g., bus) interfaces	Same as Event Source

Table 6.1 *Task-identification strategies (continued)*

Strategy	Description	Pros	Cons
Recurrence Properties	Group together events of similar recurrence properties such as period, minimum inter-arrival time, or deadline	Best for schedulability; can lead to optimal scheduling in terms of timely response to events	More complex
Safety Assurance	Group special safety and reliability assurance behaviors together (e.g., watchdogs)	Good way to add on safety and reliability processing for safety-critical and high-reliability systems	Suboptimal performance

It is common to mix multiple strategies in the same system. For example, you have use multitasking preemption based on recurrence properties as the primary scheduling strategy and also have interrupt handling for highly urgent event handling.

Another important activity is to identify the shared resources and their methods of access. A shared resource is an entity, normally modeled as a class or object (depending on whether you're referring to a resource type or resource instance) that provides a quantifiably finite set of services or data. Normally resources must be accessed by one client at a time, through a process known as serialization. This can be done with a variety of different resource-sharing patterns[8] such as the Guarded Call (i.e., using a mutex semaphore), Critical Section (i.e., turning task switching off during the use of the resource), or Message Queuing (i.e., queuing asynchronous access requests and handling them on a first-come–first-served basis).

Another issue for the concurrency architecture is to identify the scheduling patterns used. The scheduling pattern determines which tasks will run under what circumstances. The simplest is the Cyclic Executive pattern, in which each task runs in a sequence; once the task set has run through, the scheduler starts back with the first task. While this approach has the benefit of simplicity, it is demonstrably suboptimal in terms of responsiveness to incoming events; further, it is difficult to share data or synchronize with other tasks, and each task must complete before the next task can begin. A round robin pattern scheduler is only a bit more complex. In a round robin scheduler, the tasks need not complete, but must reach points at

---

8   For a detailed description of these patterns, see the author's *Real-Time Design Patterns: Robust Scalable Architectures for Real-Time Systems*, Addison-Wesley 2002, Chapters 5–7.

which they voluntarily relinquish control back to the scheduler to let other tasks run. A time multiplexed scheduler (also known as "Timed Round Robin" or TDMA scheduler) uses a time-based interrupt to stop execution of a task and begin the next in the sequence. These last three scheduling patterns are all "fairness doctrine" patterns, meaning that every task gets an equal chance to run and so they don't contain any of the notions of priority, urgency, or criticality.

There are a number of patterns that support these notions—notably the Static Priority Pattern and the Dynamic Priority Pattern. In these patterns, each task is assigned a numeric value, called priority, that determines when the task will run. When a set of tasks is ready to run, the task with the highest priority runs. When it completes, the waiting task with the next highest priority runs. If, while a task is running, a higher priority task becomes ready to run, then the scheduler stops it and places it back on the ready queue, and the higher priority task runs instead. The Static Priority pattern assigns the priority at design time. A common specialized form of the Static Priority pattern is rate monotonic scheduling (RMS). In RMS, the following assumptions are made: 1) all tasks are periodic, 2) the deadline is assumed to occur at the end of the period, and 3) all tasks can be interrupted at any point. RMS scheduling then assigns the priority of the task on the basis of the period. The shorter the period, the higher the priority. RMS scheduling is both optimal and stable. By optimal, we mean that if the task set is schedulable by any pattern, then RMS can schedule it as well (that is, you can't do better). It is stable in the sense that, in an overload situation where task deadlines cannot be met, it is possible to predict which tasks will fail—the lower priority ones.

Dynamic scheduling patterns assign the task priority at run-time based on some run-time criteria. The most common strategy is called Earliest Deadline First (EDF). EDF has the same basic assumptions as RMS but assigns priority on the basis deadline—the task with the nearest deadline has the highest priority. Thus, when a task becomes ready to run, its next deadline is calculated and then its priority is assigned on the basis of which tasks have nearer or farther deadlines. EDF is optimal but not stable; because deadlines are computed during run-time, it is not possible to predict which tasks will fail in an overload situation.

All these scheduling approaches assume task independence but have been extended to support the inevitable cases in which tasks are not, in fact, truly independent. In actual systems, most tasks must synchronize their execution at explicit synchronization points or share data and services from resources.

The last major concern in the concurrency architecture is to identify the task synchronization points and decide how to perform the synchronization. It's important to understand the basic independence-of-execution semantics of task execution. Each task can be thought of as a fully ordered series of actions. The whole point of using tasks is that the relative order of execution of the action sequences in the tasks is unimportant to the correct execution of the other tasks. The other tasks neither know, nor care, where some other task is in its sequence of actions. If the actions between two different tasks are tightly coupled, then they should not be modeled as independent tasks in the first place.

That having been said, in most systems, tasks do depend on the order of execution of other tasks but usually only in limited and well-defined ways. Consider Figure 6.6. In this figure we use an activity diagram to show the overall process for making coffee. Between the fork and the join, we show two task threads, which are independent. One of these is focused on boiling the water while the other is focused on preparing the coffee serving. However, both these tasks must be at the proper synchronization point (i.e., the water is hot and the coffee serving is prepared) before processing (i.e., mixing the coffee into the water) can proceed. Whether the water is heated before the coffee is removed from the freezer or not doesn't matter—the order of execution of those actions is independent. However, there is an explicit

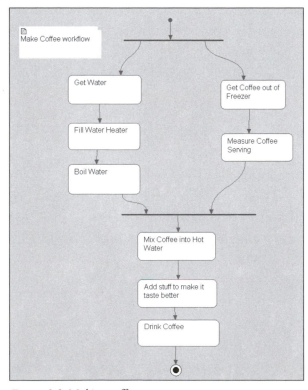

Figure 6.6 Making coffee

synchronization point at which processing cannot proceed unless both tasks are at a certain point in their execution sequences.

Let us now consider modeling the concurrency architecture of a simple example. Figure 6.7 is a simple control system. At the top left of the figure, the environmental data is acquired, filtered and averaged from four different sensors. The data is acquired under the control of the SensorManager at a rate of 10 Hz (i.e., 100-ms rate) and this is used to update the AverageSensedData. The Controller oversees the entire system and every 30 seconds it may choose to adjust the control points for the high-speed closed loop controller of the actuators based upon the averaged data. The high-speed closed loop control updates the output at 100 Hz (i.e., 10-ms rate) and does high-speed control to reduce the error between the monitored average values and the control set points. So how do we turn the analysis model in Figure 6.7 into a design model with concurrency?

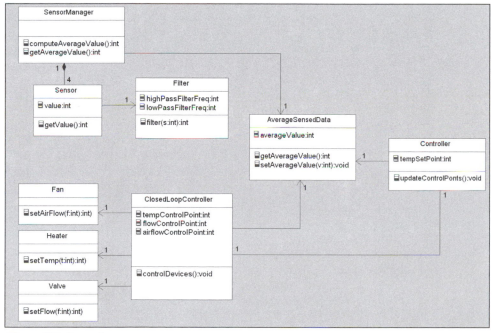

Figure 6.7 Control system

In this example, we'll use the recurrence properties to determine the task set. Actions occur at three different periods: 10 updates/sec for data acquisition, 100 updates/sec for closed loop control, and one update every 30 seconds for the update of the set points. Each of these processing sets will become an «active» object managing its own task. The AverageSensedData class is a resource shared among the task threads. For the resource sharing policy, we'll use the Guarded Call pattern to enforce serialization of the access to the data.[9]

To make this into a task diagram,[10] these classes must be made parts of the relevant «active» objects.[11] Lastly, add ports to connect the objects across the thread boundaries.

An idiosyncrasy of the UML is that you cannot connect ports of classes with associations; you can only connect ports on instances with links. Thus, in the resulting task diagram of Figure 6.8, we've made the tasks instances and connected those instances with links to show the connections. We've also added the period and priority in constraints anchored to the appropriate «active» objects.[12] To be complete, we also added the required interface (iAveragedData) to specify the ports.[13]

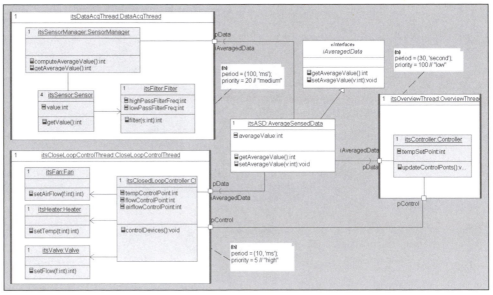

Figure 6.8 Concurrency design model

9   In Rhapsody, this is very straightforward. Simply set the concurrency property of the relevant operations from "Sequential" to "Guarded" and Rhapsody will automatically create a semaphore to manage the serialization of access to the methods.

10 A task diagram is just a usage of a class or object diagram—i.e., its mission is to depict the concurrency architecture.

11 In Rhapsody this can be done by first showing the structured view of the active class, then dragging the analysis class inside. Once there, right-click on the class you want to make a part and select "Make an Object." Make the encapsulating class active by setting its concurrency property to "active."

12 In this case, you will note that the larger the value of the priority property, the lower its priority. This is operating-system specific.

13 The observant reader may notice that we use ports only in a very limited way in the example. Ports are a design pattern and have costs as well as benefits. We tend to use them only between architectural objects that need to delegate to internal parts.

Concurrency architecture can be highly complex and the interested reader is referred elsewhere for more details on the ins and outs.[14] For the purpose of this text, we will select the strategies for you to enter into your design. Nevertheless, it is useful to have a quick glossary of terms important in the concurrency and resource model:

Table 6.2 Some concurrency definitions

Term	Definition
Arrival pattern	The recurrence property that specifies when events of a given type occur: e.g., periodic or aperiodic. Arrival patterns are detailed with quantitative information such as the period, jitter, minimum inter-arrival time, and/or average interarrival time
Blocking time	The amount of time a high-priority task is prevented from execution because a lower-priority task owns a needed resource
Criticality	The importance of the completion of an action
Deadline	In hard real-time systems, the time after an initiating event that the action must be completed to ensure correct system behavior
Deadlock	A condition in which a system is waiting for a condition that can never, in principle, occur. Can only occur when four conditions are met: 1) tasks can be preempted, 2) resources can be locked while waiting for others, 3) tasks can suspend when owning resources, and 4) a circular waiting condition exists
Execution time	The amount of time required to execute the response to an initiating event
Hard real-time	A real-time system characterized by the absolute need to adhere to a set of deadlines
Interarrival time	The time between the arrival of events of the same type; for periodic tasks this is a constant, but for aperiodic events it can vary widely
Jitter	The variation in the actual arrival time of periodic events
Period	The length of time between the arrival of periodic events
Priority	In a multitasking preemptive system, the priority is a numeric value that is used to select which task, from a set of tasks currently ready to run, will run preferentially
Race condition	A condition in which the computational result depends on a sequence of actions, but whose order is inherently unknown or unknowable

---

[14] See, for example, *Real-Time UML: Advances in the UML for Real-Time Systems*, Addison-Wesley, 2004 or *Doing Hard Time: Developing Real-Time Systems with UML, Objects, Frameworks, and Patterns*, Addison-Wesley, 1999, both by the author.

Table 6.2 Some concurrency definitions (continued)

Term	Definition
Real-time	A system in which the specification of correctness includes a timeliness measure
Resource	An element which provides information or services of a quantifiably finite nature—e.g., a set of services which must be invoked atomically or provides a set of objects from a finite pool
Schedulability	The mathematically demonstrated ability for a set of tasks in a hard real-time system to always meet their deadlines
Synchronization pattern	The means by which tasks will synchronize their execution at specific, defined points
Timeliness	The ability of a task to always meet its deadlines
Urgency	The immediacy of the need to handle an event or to complete an action

Now on to the main event …

For the Roadrunner Traffic Light Control System, we begin this process with the collaborations constructed from the object analysis model. The collaboration we will work with for this exercise is shown in Figure 6.9.

For this problem, take the collaboration shown in Figure 6.9 and make this into a design-level collaboration with concurrency. Remember to apply the task identification strategies (that's why they're there!); walk though the concurrency architecture workflow and make sure you either do all the steps or have convinced yourself that you don't need to. Decide what scheduling pattern you'd like to use, whether it's cyclic executive, static priority, or whatever. Normally, the scheduling pattern is not explicitly shown in the task diagram because it is relegated to the underlying computing-platform operating system.

For the Coyote UAV, we will continue to focus on image acquisition. The collaboration to add concurrency is shown in Figure 6.10.

Apply the task-identification strategies and the concurrency architecture workflow to create a task diagram for this collaboration. Also, decide on a scheduling strategy.

For both solutions, justify your selection of a scheduling pattern. How would the behavior be different if a different strategy was selected? What would the pros and cons of the various strategies be?

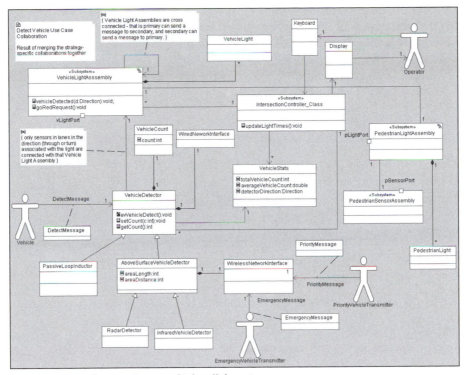

Figure 6.9 Roadrunner Detect vehicle collaboration

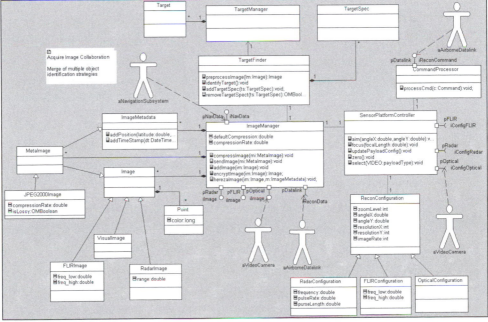

Figure 6.10 Acquire Image collaboration

Problem 6.2   Distribution Architecture

The distribution architecture focuses on how objects are allocated to different processors and the means by which they collaborate. These means include not only the underlying communications protocols but also the policies and patterns that govern their use.

Distribution patterns come in two basic flavors: asymmetric and symmetric. Asymmetric architectures are distinguished in that the allocation of objects to processing nodes is done at design time. This enables a number of simplifying assumptions to be made for the collaboration, since client objects can reliably know where their servers are and how to talk with them. Symmetric architectures are characterized by run-time allocation of objects to processor nodes. This organization is more flexible than asymmetric but comes at a cost of increased complexity, required infrastructure support, and possibly run-time overhead. Another benefit of symmetric architectures can be improved reliability, because objects can be relocated when a processor node fails.

Asymmetric architectures can be supported by simple infrastructures such as networks (e.g., Ethernet), busses (e.g., CAN or 1553 bus), serial links (e.g., USB or RS-232), or shared memory (e.g., dual-ported RAM). In all these cases, it is most common to have a general-purpose communication protocol (i.e., one that is not application-specific) and use this protocol to exchange application data and request services. A communication protocol is generally a logically multitiered structure in which low-level bit or byte exchange facilitates the logical exchange of application messages at a higher level of abstraction in the protocol. For example, Ethernet can provide the underlying data-link and transport facilities for TCP datagrams, but a system can implement the higher-level exchange mechanism of Remote Procedure Calls (RPCs) using these lower-level facilities. The exchange of messages between remote applications takes place at the application layer in the protocol stack, but this is ultimately realized by lower-level mechanisms in the transport, data link, or physical protocol layers.

It is important, for good system design, to isolate the protocol details away from the application semantics as much as possible. This not only simplifies the application but also allows the application to run more easily over different communication protocols. Distribution patterns appear at or just above the application layer in the protocol stack. As with all patterns, distribution patterns optimize some aspects of the system design at the expense of others.

By far, asymmetric architectures are more usual in embedded systems. They are simpler and tend to be more time-efficient than symmetric architectures, which fits in well with the highly constrained environments in which many embedded systems must reside. However, the simple approaches often don't scale well to larger-scale systems. The most common infrastructure for use in symmetric systems is to use an ORB (object request broker) architecture pattern, and the most common ORB standard is CORBA (common object request broker architecture).[15]

The Roadrunner Traffic Light Control System is a moderately distributed system. Sensors and actuators are scattered about the intersection and may have small processors that allow them to communicate with the intersection controller. Certainly a connection must exist between the front panel display and the intersection controller as well. The system must also connect to the remote service and management system via a wired Ethernet connection and also must support simple communication with high-priority and emergency vehicles. To construct the distribution model for the Roadrunner Traffic Light Control System, take the collaboration[16] from the object-analysis model and add objects to facilitate communications, such as protocol stacks, message types, and distribution pattern-specific classes and add them to the collaboration.

For this problem, make the following assumptions:

- The distribution architecture is asymmetric

- Each light assembly and sensor assembly will be connected to the Intersection controller and each other via a serial multidrop RS-422 link. Simple messages shall be exchanged using synchronous RPCs (remote procedure calls) over the serial link, including:

  - setLightMode(ModeID:int)

  - herezaEvent(eventSource:int; ev:EventType) // for collaboration among lights and sensors

  - detectEvent(sourceID:int)

- The multidrop RS-422 link is managed as a token ring, each node getting the master token for a brief period (not to exceed 20ms) to send messages. This ensures that the entire ring is cycled in no less than 1/5 second.

---

[15] See *Pattern-Oriented Software Architecture, Volume 2, Patterns for Concurrent and Networked Objects* by Doug Schmidt, Michael Stal, Hans Rohnert, and Frank Buschmann, John Wiley and Sons, 2000, for a detailed description of a number of different CORBA-related patterns.

[16] You can omit the threads done for the concurrency model. They'll still be there in the system, but now we're focusing on the communication aspects, not the concurrency aspects, so we will omit the latter. This is in keeping with the "single importance concept per diagram" guideline.

- The connection to the Front Panel Display shall be an Ethernet link.

- The connection to the remote service and management system shall be over an Ethernet link

- The above-surface vehicle detectors support both wired (multidrop RS-422 above) and wireless communications. The wireless protocol shall be the same as used for the transceiver used to detect high-priority and emergency vehicles.

- The transceiver shall support a simple, custom protocol that is focused and directional, allowing the reception (only) of the following messages

  - HighPriorityVehicleApproaching

  - EmergencyVehicleApproaching

- There shall be four IR transceivers (one for each direction) that transmit the emergency and priority vehicle messages to the Intersection Controller.

For the Coyote UAV, we'll focus on adding the distribution architecture to the Reconnaissance Management subsystem but a brief description of the overall aircraft communication architecture is appropriate. The aircraft has two telemetry links to the ground using different frequencies. The low-speed datalink is encrypted and secure with a bandwidth of 100 bits per second (bit/s). The high-speed datalink is unencrypted and is used for high-bandwidth video and image streaming. It must support 640 × 480 resolution images at 30 fps. If uncompressed, that means that it must support a bandwidth of 640 × 480 × 10 × 30 = 92,160,000 bit/s. However, lossless compression of at least 30% on average is possible, leading to a required minimum bandwidth of 64,512,000 bit/s. Since we need some margin for error, a 100 Mbit/s should be adequate to meet the needs. That means that the internal network must be able to handle this bandwidth in addition to other data necessary to fly and maneuver the aircraft; a minimum of 200 Mbit/s of real bandwidth should be more than adequate. If we use a CDMA (collision detect multiple access) protocol, such as Ethernet, then such a network saturates at about 30% of available bandwidth. Thus, we need a 1-Gbit/s Ethernet internal bus to ensure we have no significant delay in the delivery of the data due to collisions on the network if everything operates on the same network.[17]

How best to model this? As always, there are multiple approaches to solving the problem. The simplest, with Rhapsody, is to simply stereotype the interfaces that specify

---

[17] If the 1-Gbit/s network is unpalatable, then you might decide to either run multiple networks (one for command and control and another for high-speed data), or reduce the fidelity of the video.

the contracts on the ports «CORBAInterface». Rhapsody then "CORBA-enables" these ports so that it generates all the IDL for the proxies that marshal and unmarshal the services invoked across those ports. But what if we don't want to use CORBA?

In some sense the most ideal solution would be to create a class called TCP-PortClass, that has the structure, relations (to the protocol stack, for example), and behavior to marshal and unmarshal service requests and responses and then send them via datagrams, with sessions and sockets, as appropriate. Then use instances of this class to replace the existing ports on the subsystem and other structured classes. This, however, is not easily supported in modeling tools since in most tools ports are not easily extensible, requiring you to write significant amounts of code to implement the solution.

Another, less satisfactory solution from the modeling point of view, is to refactor the architecture, deleting the old "semantic" ports that are specified in terms of application data and services, and replace them with TCP/IP ports that are specified in terms of their ability to receive (or require) requests for sessions, sockets, and that can send and receive datagrams. This approach is less satisfactory from a model standpoint because it disrupts the continuity from the original architecture model—which is based on data and services in the system—and modifies it to be based on the design infrastructure rather than the application semantics.

In order to make the best choice, we need to ask ourselves "What do we want from our solution?" An ideal solution should:

- Provide the connections to the other systems over the desired network
- Require minimal modification (in the best case, no modification) to the existing collaborations and to the existing architecture
- Be easy to reapply to different distribution architectures, networks, and datalinks

One more approach, and the one recommended here, is to create proxies using a variation of the proxy pattern called the port proxy pattern. A proxy is a "stand-in" for another element and in this case understands how to marshal and unmarshal the service requests and responses into TCP/IP datagrams, set up connections and sessions, and so on. The port proxy pattern is specifically designed to address the issue of connections across ports that are to be implemented using a networking infrastructure.

The port proxy pattern problem is shown in Figure 6.11.

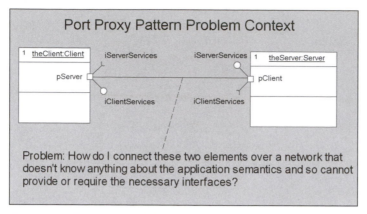

Figure 6.11 Port proxy problem

The solution—adding in proxies to manage the communication over the networking infrastructure—is shown in Figure 6.12. It is a "simple" matter of creating proxies that manage the networking communications. These proxies support the required application semantic interfaces and then create the necessary network data structures (such as datagrams) and execute the necessary network behaviors (such as creating sessions and sending and receiving network packets) to achieve the communication goals. The primary advantage of this pattern is that the original objects (in this case, the subsystems) don't need to change at all to support the network infrastructure. The change is isolated to the links among the objects; now they are connected via proxies instead of directly. The original architecture still executes because the interfaces haven't changed and the new architecture with the proxies added works as well.

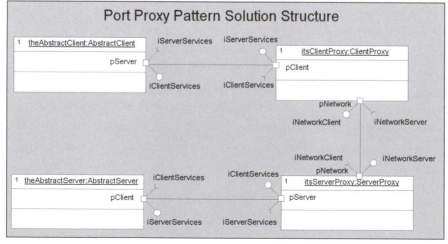

Figure 6.12 Port proxy pattern structure

Your task for the CUAV is to apply this port proxy pattern to the Reconnaissance Management, Airborne Datalink, Navigation, Targeting subsystems and payloads. Draw the related subsystems and connect them via proxies.

Problem 6.3 Safety and Reliability Architecture

The safety and reliability view specifies how the architecture identifies, isolates, and corrects faults at run-time. At the architectural level, this boils down to ways in which redundancy is defined and used to achieve the system safety and reliability goals.

Reliability is a measure of the "up-time" or "availability" of a system; specifically, it is the probability that a computation will successfully complete before the system fails. It is normally estimated with mean time between failure, or MTBF. MTBF is a statistical estimate of the probability of failure, and applies to stochastic failure modes.

Reducing the system down time increases reliability by improving the MTBF. Redundancy is one design approach that increases availability because, if one component fails, another takes its place. Of course, redundancy only improves reliability when the failures of the redundant components are independent.[18] The situation in which a single failure can bring down multiple components is called a common mode failure. One example of a common-mode failure is running software for both the primary and secondary processing on the same CPU. Should the processor fail, then both components will fail. In reliability analysis, great care must be taken to avoid common mode failures or to provide additional redundancy in the event of an element, common to all redundant components, failing.

The reliability of a component does not depend upon what happens after the system fails. That is, regardless of what happens after the failure, the reliability of the system remains the same. Clearly the primary concern relative to the reliability of a system is the availability of its functions to the user.

Safety is very different from reliability, but a great deal of analysis affects both safety and reliability. A safe system is one that does not incur too much risk of loss, either to persons or equipment. A hazard is an undesirable event or condition that can occur during system operation. Risk is a quantitative measure of how dangerous a system is and is usually specified as:

$$Risk = Hazard_{severity} * Hazard_{likelihood}$$

---

[18] Strict independence isn't required to have a beneficial effect. Weakly correlated failure modes still offer improved tolerance to faults over tightly correlated failure modes.

The failure of a jet engine is unlikely, but the consequences can be very high. Overall, the risk of flying in a plane is tolerable because, even though it is unlikely that you would survive a crash from 30,000 feet, such an incident is an extremely unlikely. At the other end of the spectrum, there are events that are common, but are of lesser concern. A battery-operated radio has a hazard of electric shock but the risk is acceptable because even though the likelihood of the hazard manifestation is relatively high, its severity is low.[19]

Faults come in two flavors. Errors are systematic faults introduced in analysis, design, implementation, or deployment. By "systematic," we mean that the error is always present, even though it may not be always be manifest. In contrast, failures are random faults that occur when something breaks. Hardware exhibits both errors and failures but software exhibits only errors. The distinction between error and failure is important because different design patterns optimize the system against these concerns differently.

The key to managing both safety and reliability is redundancy. Redundancy improves reliability because it allows the system to continue to work in the presence of faults. Simply, the redundant system elements can take over the functionality of broken ones and continue to provide system functionality. For improving safety, additional elements are needed to monitor the system to ensure that it is operating properly and possibly other elements are needed to either shut down the system in a safe way or take over the required functionality. The goal of redundancy used for safety is different than for reliability. The concern is not about continuing to provide functionality, but instead to ensure that there is no loss (to either persons or equipment).

The example I like to use to demonstrate the difference is the handgun versus my ancient Plymouth station wagon. The handgun is a highly reliable piece of equipment—most of them fire when dirty or even under water. It is, however, patently not very safe since, even in the absence of a fault, you can (and people do) shoot yourself in the foot. On the other hand, my enormous 1972-vintage station wagon (affectionately referred to as "The Hulk") is the safest automobile on the planet. It has a fail-safe state[20] ("OFF") and it spends all of its time in that state. So while the vehicle is very safe, it is not at all reliable.

---

[19] For more on safety, see *Doing Hard Time: Developing Real-Time Systems with UML, Objects, Frameworks, and Patterns*, Addison-Wesley, 1999 by Bruce Powel Douglass or Safety Critical Computer Systems by Neil Storey, Addison-Wesley, 1996. For a general overview of the issues of safety in software-intensive systems, see *Safeware: System Safety and Computers* by Nancy Leveson, Addison-Wesley, 1995.

[20] A fail-safe state is a condition (state) of a system known to be always safe, i.e., free from loss.

As with the other architectural dimensions, safety and reliability are achieved through the application of architectural design patterns.[21] All design patterns have costs and benefits, and selecting good safety patterns requires balancing the design concerns, such as:

- Development cost

- Recurring (manufacturing) cost

- Level of safety needed

- Level of reliability needed

- Coverage of systematic faults (errors)

- Coverage of random faults (failures)

- Complexity

- Resource demand

- Ease of certification against relevant standards

In general, safety and reliability patterns can be categorized into either homogeneous or heterogeneous patterns. The former creates exact replicas of the architectural elements to provide redundant processing, and adds glue logic to determine when and under what circumstances the replicas run. The latter patterns use different implementations, designs, or approaches to provide redundant processing. These systems can be further subdivided into lightweight or heavyweight patterns. Lightweight patterns use fewer resources but may not be able to provide the full functionality or fidelity of the primary system elements. Heavyweight redundancy replicates the full functionality but at a greater cost.

Assessing the adequacy of the design solution is done with safety analysis (such as Fault Tree Analysis [FTA[22]]) and/or reliability analysis (such as Failure Mode and Effect Analysis [FMEA[23]]). FTA uses logical operators to connection conditions (some of which may be undesirable) with events, such as failures. The notational elements of the FTA are shown in Figure 6.13.

---

[21] For details of safety-related patterns, see the author's *Real-Time Design Patterns: Robust Scalable Architectures for Real-Time Systems*, Addison-Wesley, 2002, Chapter 9.

[22] An excellent resource is *The Fault Tree Handbook*, NUREG 0492 (Nuclear Regulatory Commission, 1981), available at *http://www.nrc.gov/reading-rm/doc-collections/nuregs/staff/sr0492/*.

[23] See *http://www.fmeainfocentre.com/* for some FMEA resources.

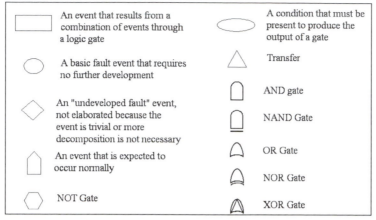

Figure 6.13 FTA notational elements

The purpose is to identify the conditions and events that must occur for hazard-ous conditions to arise. The logical operators are AND, NAND (Not AND), OR, NOR (Not OR), XOR, and NOT. The circle is a basic fault and the rectangle is an "intermediate fault"—that is, a result of a logic operator working on more primitive faults. A typical if simplified FTA is shown in Figure 6.14. The system depicted is a

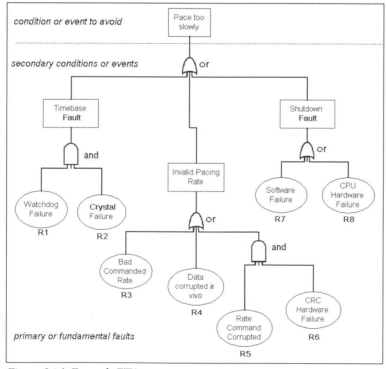

Figure 6.14 Example FTA

cardiac pacemaker, and the hazardous condition depicted is Pacing Too Slowly. Once such an analysis has been performed, safety or control measures are added that must be ANDed with the other conditions to allow the hazardous condition to arise. In the example FTA shown, for a Timebase Fault to arise, both the CPU crystal must fail AND the added watchdog safety measure must fail. I typically do a separate FTA for each identified hazardous condition when doing a system safety analysis.

The result of this analysis usually culminates with a hazard analysis that specifies each fault with a description, severity, likelihood, risk, fault tolerance time, control measure (design approach to handle the fault), fault detection time, and fault exposure time. I represent this information as a spreadsheet, such as that shown in Figure 6.15, which shows some hazards for a patient ventilator. The actual performance of safety analysis is beyond the scope of this book, but interested readers are referred to the cited references for more information.

Safety and reliability analyses are not a part of the UML, but are used in conjunction with the UML to assess the safety and reliability of the systems specified in the UML.

Most of the commonly used safety and reliability patterns are based on the channel pattern. This pattern is shown in Figure 6.16. A channel is a kind of subsystem that performs "end-to-end" functionality; that is, it takes data from physical sensors and performs a series of data transformational steps, culminating with output actuator control. The advantage of this organization is that it is easy (although not always inexpensive!) to replicate the entire channel so that if any point in the channel fails, another channel can take over the required functionality.

Figure 6.17 shows an example simplified from a medical delivery system. The channel shown delivers oxygen. The actors (shown as objects with the «actorInstance» stereotype) connect to the channels. Note that different O2 sensors and gas mixers connect to the different channels, as required to avoid common mode failures. Inside the channels, the acquired data goes through a series of transformations, including a control loop, until the appropriate settings for the gas mixer can be computed and then sent to the gas mixer actor. Additionally, there is a monitor object that compares the output of the channel with the input (and, presumably, the desired control settings) and can switch over to the other channel if the first fails.

This pattern is single-point failure safe. If any element in the first channel fails (including the sensor or gas mixer), the monitor detects it and switches over to the other channel. If the monitor fails and switches to the other channel inappropriately, then the other channel still works properly (since in single-point failure safety analysis, we assume only a single fault), and the proper amount of oxygen will still be delivered.

Hazard Analysis

How to use this spreadsheet

The hazard analysis spreadsheet computes risk = severity * likelihood, where severity is a ranking of 1 (very low) to 10 (very high).

Note that various safety standards may use a different range of severity. Likelihood is the probability of occurrence of the hazard in the life expectancy of the product (0.0 to 1.0). Risk is computed from these values.

Exposure time is computed as the sum of the Detection Time + Action Time. For a safe system this value must be less than the Tolerance time.

Is Safe is computed as = Exposure Time <= Tolerance Time.

Note that the spreadsheet assumes that the time units are the same for an entire row.

Hazard	Fault	Severity (1 [low] – 10 [high])	Likelihood (0.0 – 1.0)	Computed Risk	Time units	Tolerance Time	Detection Time	Control Measure	Control Action Time	Exposure Time	Is Safe?
Hypoventilation	Breathing tube disconnect	10	0.02	0.2	minutes	5	0.5	Blood oxygen sensor	2	2.5	TRUE
Hypoventilation	Ventilator timer error	10	0.02	0.2	minutes	5	0.5	Independent pressure sensor with alarming	2	2.5	TRUE
Hypoventilation	Gas Supply Failure	10	0.04	0.4	minutes	5	0.05	Ventilator incoming gas pressure sensor	2	2.05	TRUE
Hypoxia	Gas mixer failure	10	0.06	0.6	minutes	5	0.05	Inspiratory limb 02 sensor	2	2.05	TRUE
Hyperventilation	Ventilator timer error	8	0.01	0.08	minutes	20	0.5	Blood oxygen sensor	2	2.5	TRUE
Overpressure	Pump failure; expiratory tube blockage	10	0.03	0.3	ms	200	10	Secondary pressure sensor with auto release valve	5	15	TRUE

Figure 6.15 Hazard analysis

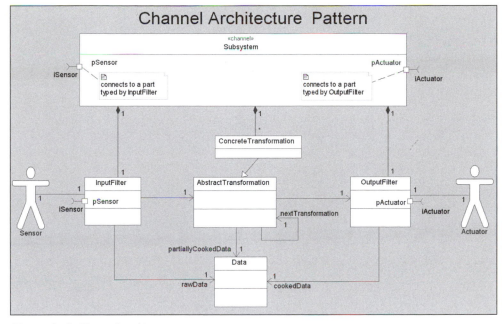

Figure 6.16 Channel architecture pattern

This pattern, using homogeneous redundancy, has the benefits of single-point failure safety, low design cost, and it can continue to provide services safely in the presence of a failure. However, it cannot handle errors (since any errors will appear in both channels) and it has a high recurring cost. The first problem (handling errors) can be mitigated by using heterogeneous (aka "diverse") channels—using a different design or design team—but with a higher design cost.

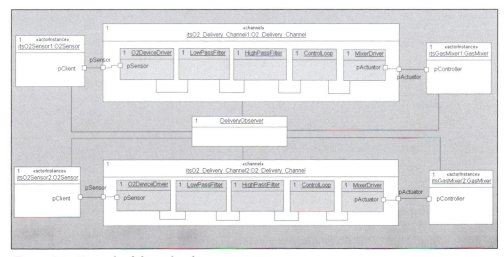

Figure 6.17 Example of channel architecture pattern

Another common pattern is the monitor-actuator pattern. This is a variant of the channel pattern that has relatively low design and recurring costs but cannot continue in the presence of faults; that is, it requires that the system have a fail-safe state. The pattern structure is shown in Figure 6.18.

The monitor-actuator pattern works by having a separate monitor that monitors the physical environment. If the actuation channel isn't having the desired effect or the results are not within nominal operating conditions, then the monitor commands the channel into its fail-safe state (usually off or manual control). If monitor has a fault, then either it allows the actuation channel to continue (which, by single-point failure safety rules is still operating properly) or it mistakenly reverts to the fail-safe state. In all cases, if there are no common-mode faults, the system remains safe. Note that to avoid common mode faults, the monitor must use a different sensor than the actuation channel.

It should be noted that the monitor's computation of "correct" may be complex. There is normally some range variance for a variety of reasons, and often flow or diffusion dynamics must be part of the computation. For example, in an oxygen-flow system, the O2 flow at the point of measurement won't change the same instant the set point is changed. Partial differential equations can be used to estimate the expected rate of change as a function of the volume of the breathing circuit, the flow rates, predicted nonlaminar flow, diffusion, and O2 update (extraction) by the patient.

Figure 6.19 shows a straightforward example of the monitor-actuator pattern, applied to the same system as the previous pattern. In this case, a separate sensor, perhaps connected at the terminal (patient) end of the inspiratory breathing tube, monitors delivered oxygen flow. The monitor compares it to the desired flow from the control panel (ultimately set by the physician). If it is not correct, then it shuts down the delivery of O2 and alarms to the attending physician, who (as part of the "safety loop") then fixes the problem and restores the proper oxygen flow to the patient.

The last safety and reliability pattern we'll mention in this chapter is the single channel protected pattern. This pattern offers both low design and recurring costs but provides less coverage than some of the other patterns. It is similar to the monitor actuator pattern in that it may only be used in safety-critical systems if there is a fail-safe state.

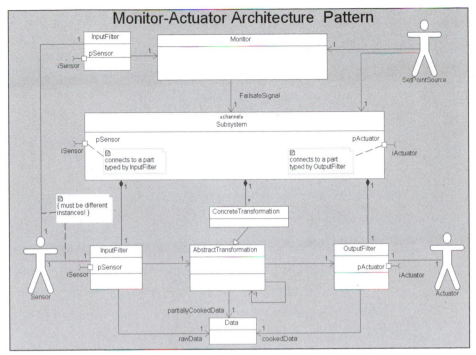

Figure 6.18 Monitor actuator pattern

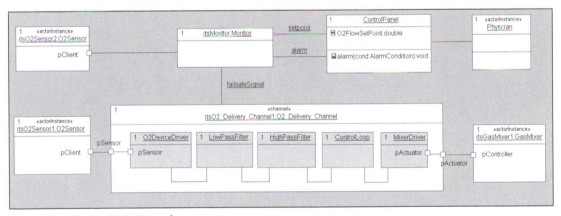

Figure 6.19 Example monitor actuator pattern

The structure of the pattern is shown in Figure 6.20. You can see clearly that it is a variant of the channel pattern. It specializes that pattern by providing lightweight in-line data validity checks. It can detect data corruption and some computational error but, unlike the monitor-actuator pattern, does not compare the physical output with the expected result. Internal data- and processing-specific safety checks are typically things like:

- Data corruption (feed-forward) checks
  - CRC
  - Parity check (weak)
  - Hamming codes
  - Redundant data check
  - Inverted redundant data check (e.g., store the date in a one's complement (bit inversion) form and when reading the data, invert the copy and compare to the stored value).
- Reasonableness checks
  - Range checks (e.g., patient weight should be in the range of 0–400 kg)
  - A priori rule checks, e.g.,
    - height should be a function of range of weight with a range around a mean
    - if patient type is neonate, then weight range is 0–15 kg
    - if patient type is pediatric, then weight range is 10 kg–100 kg
  - Unit checks (e.g., kilograms, grams, pounds, when data may come from sources with different units)
- Backward derivation (feedback) checks
  - Run the computation backwards to compute the original value and compare

It is also common to run background tasks to detect stack overflow, memory faults, and liveness (e.g., lack of deadlock). Such checks are commonly known as "built-in tests."

The channel provides a very low-cost but less-complete check. An example is shown in Figure 6.21.

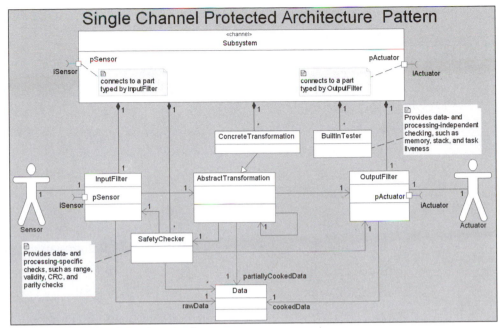

Figure 6.20 Single-channel protected pattern

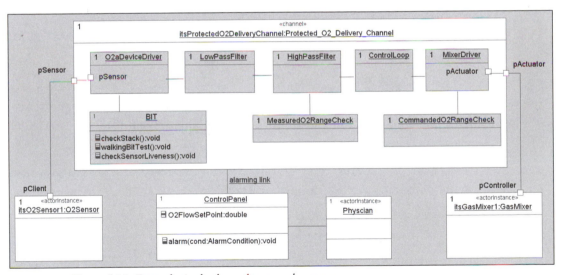

Figure 6.21 Example single-channel protected pattern

Both the example systems in this book are safety-critical. The Roadrunner's reason for existence is to provide a safe way for traffic to flow at points that are inherently dangerous (intersections). This traffic flow includes both vehicles and pedestrians. The system works by basically making everyone take turns so that they proceed through the intersection only when it is safe to do so. There are faults that could affect the safe fulfillment of this functionality, such as:

- Both primary and secondary through traffic have GREEN lights
- Pedestrian traffic has a WALK light at the same time that the orthogonal vehicle traffic has a GREEN light
- The turn lanes in SIM (simultaneous mode) are green at the same time the through traffic in the same orientation has GREEN
- The turn lanes in the SEQ (sequential mode) are green when the opposing traffic has a green light
- All lights are off

The same rules may be applied to yellow lights and flashing wait as well, since traffic normally continues to flow under these conditions, at least for a while.

For the first part of this exercise:

- Take the collaboration done previously (shown again as Figure 6.22) and do a separate FTA on each of the fault situations below:
  - Both primary and secondary through traffic have GREEN lights
  - Pedestrian traffic has a WALK light at the same time that the orthogonal vehicle traffic has a GREEN light
  - All lights are off
- From the result of this analysis, create a hazard analysis in which hazards and faults are identified and control measures are added, including estimated fault tolerance times
- Determine how architectural redundancy should be added to make the system single-point fault safe for the analyzed fault situations
- Draw the safety architecture class or object diagram showing how the elements collaborate to achieve the desired safety.

For the second part of this problem, let's consider the CUAV safety requirements. The major safety risks are:

- Loss of the aircraft
- Loss of life (friendly or noncombatant) due to crashing the aircraft
- Loss of life (friendly or noncombatant) because of a misdirected missile (from the CUAV itself) or remote fire based on data from this system (e.g., a tank

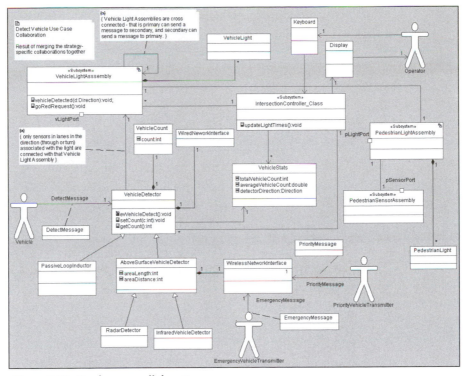

Figure 6.22 Roadrunner collaboration

receiving information from the CUAV via the ground station), or not firing on hostiles when appropriate. These hazards could arise from

– Data faults

  ■ target misidentification (i.e., "false positive")

  ■ lack of (valid) target identification (i.e., "false negative")

  ■ loss or corruption of target information either in storage or transmission, e.g.,

    – incorrect target classification[24] or Combat ID (CID)[25]

    – incorrect location or coordinates

    – loss of information about target

    – incorrect target kinematics (vector, flight path, etc.)

---

[24] Target classification is in terms of the type of platform (e.g., tank, truck, aircraft, platoon, etc.), class, unit, and/or nationality.

[25] CID is the identification of the target friendliness, such as Friend, Assumed Friend, Hostile, Suspect, Neutral, or Unknown.

- incorrect target management (e.g., failure in sensor fusion)
  - Missile faults
    - missile flight control failure
    - misfire of the missile
    - firing a missile without proper authorization
    - lost or corrupted commands or data during message transfer

The last of these is mitigated somewhat by including a mission ops person (human) in the loop and not leaving the use of lethal force entire up to automated decision processes. However, even the smartest mission ops person can be misled by getting an incorrect Combat ID in the "fog of war."

Since we've focused previously on developing the reconnaissance management subsystem, involved in both target identification, tracking, and targeting, we'll continue in that vein.

- Create an FTA for the reconnaissance management subsystem that identifies the safety hazards and risks associated with this subsystem, one per hazard.

- From the result of this analysis, create a hazard analysis in which hazards and faults are identified and control measures are added, including estimated fault tolerance times.

- Determine how architectural redundancy should be added to make the system single-point fault safe for the analyzed fault situations.

- Draw the safety architecture class or object diagram for the subsystem showing how the elements collaborate to achieve the desired safety.

Looking Ahead

We've seen that architecture is all about taking an analysis model, or platform independent model (PIM) and performing system-wide gross-level optimization. In the Harmony process, there are at least five independent but interacting areas in which these architectural design decisions are made:

- Subsystem and component architecture
- Concurrency and resource management architecture
- Distribution architecture
- Safety and reliability architecture
- Deployment architecture
- Security architecture (optional)

Architecture was categorized in this way because each of these areas has its own rich set of independent design patterns and solutions. The overall architecture is the sum of one or more patterns from each of these architectural areas of concern.

Design—including architectural design—is all about optimization. As such, it is imperative to first identify the design criteria and then rank them in order of criticality. This is because many of the design criteria (such as worst-case execution time and memory utilization) are traded off against each other by the design patterns. The ranking allows us to clearly identify and optimize the design criteria that are most important to us.

Design is discussed in terms of patterns because patterns are a way of codifying and reusing design solutions that apply to problems that reoccur in different contexts. Patterns, then, are generalized solutions of commonly occurring problems. In this chapter, we've seen how to apply design patterns in these architectural areas of concern, and just a few of the patterns that are available for use.

In the next chapter, we'll look at the next two levels down in design abstraction, mechanistic and detailed design. Design solutions in the mechanistic level of abstraction have a more limited scope—a collaboration realizing a single use case. Although not as "strategic" in its scope, mechanistic design is still critical for good real-time and embedded systems development. Detailed design decisions optimize individual classes at a time, and so are much smaller than even mechanistic design patterns.

Mechanistic and Detailed Design

What you will learn:

- **What is "mechanistic design"?**
- **Optimizing analysis collaborations**
  - Optimizing interaction
  - Optimizing reusability
  - Optimizing safety and reliability
- **What is "detailed design"?**
- **Optimizing classes and objects**
  - Optimizing for space complexity
  - Optimizing for time complexity
  - Optimizing state behavior
  - Optimizing for reusability

Overview

In this chapter, we'll deal with the two remaining levels of design abstraction; mechanistic and detailed design. The first thing to remember is that, while analysis is about semantic correctness (hence construction of the "essential model"), design is all about optimization of that analysis model against the various design criteria. Mechanistic design takes an existing analysis-level collaboration (as discussed in Chapter 5) and, based on the weighted set of design criteria, optimizes that collaboration by applying appropriate design patterns.

Mechanistic design optimizes the model at the collaboration level. A collaboration is a set of classes and objects that work together to realize a larger-scale behavior, such as the realization of a use case. Since a typical system has one to a few dozen use cases (at the high level, at least), the scope of the design decisions made for mechanistic design is an order of magnitude smaller than for architectural design. If there are 25 use-case ("analysis-level") collaborations, then mechanistic design must be applied 25 times, once per use case. Depending on the specific needs of a specific collaboration, it will use different design patterns because the context may be different and because the design criteria are likely to be different as well.

Detailed design optimizes the system at an even lower level of abstraction. Detailed design applies idioms to optimize individual class features.[1] The scope of detailed design is about an order of magnitude smaller than mechanistic design. While this sounds daunting because of the sheer number of classes in a typical sysytem, practice has shown that most classes in a well-designed system are very straightforward and require little, if any, optimization. There is usually a small set of the classes—typically 3% to 5%—that require special effort. It is for these "special needs" classes that we devote most of our effort in detailed design.

Mechanistic Design

A collaboration is a set of classes and objects working together to realize a larger-scale purpose. A collaboration is specified in terms of specific classes playing named roles (called classifier roles) that interact to produce larger-scale behavior and functionality. Solutions to design optimization problems can usually be generalized and reused in different specific contexts. Such reusable design solutions are called design patterns. The roles define, in an abstract sense, the formal parameters for the pattern. When bound with actual parameters (the classes from your analysis collaboration), the pattern is said to be instantiated. A mechanism[2] is a type of pattern that is limited in scope to a few classes (i.e., it does not have architectural scope), but is generally applicable in many circumstances. Thus. mechanistic design is the term in the Harmony process for the application of mechanisms to produce design-level collaborations. A brief overview of design patterns is given in the Overview section of the previous chapter.

The Harmony design process is very pattern oriented. Figure 7.1 shows the mechanistic design workflow for the Harmony process. To understand the workflow, it is important to remember that the recommended lifecycle in the Harmony process

---

[1]   A class feature is an internal element of a class, such as a method, attribute, port, or state.

[2]   See Booch, Grady, James Rumbaugh, and Ivar Jacobson, *The Unified Modeling Language User's Guide*, Reading, MA: Addison Wesley Longman, 1999.

is spiral; the spiral lifecycle creates the end product as a series of progressively capable versions, called iterative prototypes.[3] Normally, only one or a small number of use cases are realized in any given spiral. As more spirals are completed, the prototype becomes increasingly complete and capable, until at last the project is complete and the prototype is shipped to the customer.

In the object-analysis phase of the spiral (see Chapter 5), one analysis-level collaboration is produced per use case, so that for any given spiral, only a small number (typically 1–3) of collaborations are generated. In mechanistic design, each of these analysis-level collaborations is converted into a design-level collaboration by applying the workflow shown in Figure 7.1. As discussed in Chapter 2, these spirals are completed frequently, with 4–6 weeks being a common timeframe for a single spiral. In addition to generating new collaborations, existing collaborations may be changed in small ways to account for the new functionality added in the current spiral. This process is known as refactoring.

Just as with architectural design, it is important to explicitly specify the design criteria so that appropriate design patterns may be selected. These criteria are ranked in order of criticality. This ranking is important because many of the design criteria will be in conflict with each other, and the proper application of design patterns requires that you optimize the most important design criteria at the expense of the least important.

Once the design criteria are identified and ranked, the mechanistic design pattern literature can be searched for patterns that perform the desired optimization. Unfortunately, most references on design patterns mention pattern pros and cons only in passing as a secondary concern. In our view, the pros and cons are the primary selection criteria to identify appropriate design patterns. That is, we don't apply patterns because they are "cool," but because they achieve the desired optimizations at an acceptable trade-off cost.

Once the design patterns are selected, they must be instantiated into the collaboration. This, as discussed in the previous chapter, is a matter of inserting the classes from the pattern into the analysis collaboration, and connecting the analysis classes to the pattern classes by either subclassing from or associating with the pattern's classes.

---

[3] By *prototype*, we do not mean something hacked together to demonstrate some concept. This proto-type contains production code, but may be incomplete.

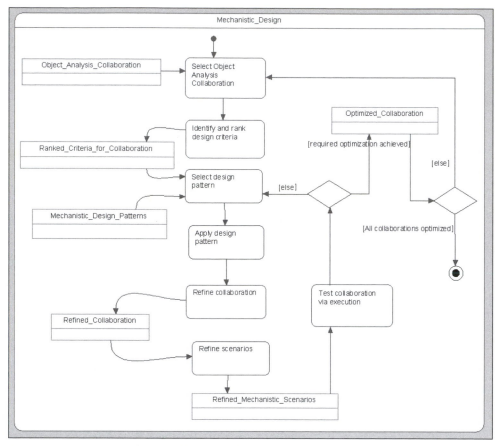

Figure 7.1 Harmony mechanistic design workflow

The classic reference on mechanistic design patterns is by Gamma, et. al.[4] It classifies patterns into three categories:

- *Creational patterns*
  These patterns deal with the object instantiation process.

- *Structural patterns*
  These patterns deal with how objects are composed into larger structures.

- *Behavioral patterns*
  These patterns deal with algorithms and assignment of responsibilities among sets of objects.

---

[4] Gamma, Helm, Johnson, and Vlissides, *Design Patterns: Elements of Reusable Object-Oriented Software*, Addison-Wesley, 1995.

This is an excellent reference and highly recommended, even though the representation of the patterns is not in the UML.

Delegation Pattern Strategy

Almost all patterns use either delegation or interface abstraction to achieve their goals. <mark>Delegation strips out some functionality from one class and places it in another.</mark> This allows the delegated class to focus on a more specific responsibility and often makes it more reusable in different contexts. The container-iterator pattern works in this way. The "whole" class owns a set of parts and must manage their containment (add, remove, find, sort, etc.). By abstracting the containment responsibility into a separate container class, the "whole" is simplified and it becomes easy to manage the parts and to replace the container type when necessary. Iterators provide "bookmarks" into the containment so that multiple clients of the collection can easily track where they are without stepping on the toes of other clients.

Figure 7.2 shows the basic structure of the container-iterator pattern.[5] Note that it uses both generalization and parameterization.[6] Figure 7.3 shows how you might

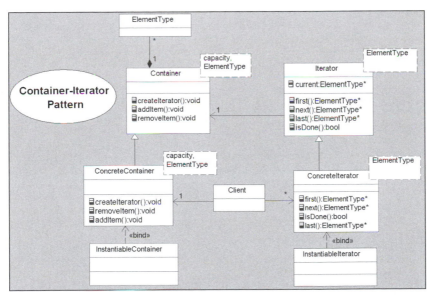

Figure 7.2 Delegation example—container-iterator pattern

---

[5]   The container-iterator pattern is so common and useful that Rhapsody provides the pattern out-of-the-box. Whenever a model contains a * multiplicity, Rhapsody (by default—you can turn it off if desired), generates containers and iterators to manage the containment.

[6]   Generalization (aka "inheritance") is used when you want to specialize the behavior but work on data of the same type. Parameterization (aka "template classes in C++ or "generics" in Ada) is used when you want to have exactly the same behavior but work on different data types.

go from an analysis class that must manage the containment of many messages to the separation of that concern into a separate queue class. Lastly, Figure 7.4 shows the details of how the pattern is instantiated to achieve this separation of concerns.

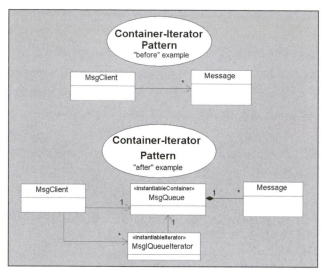

Figure 7.3 Use of container-iterator pattern

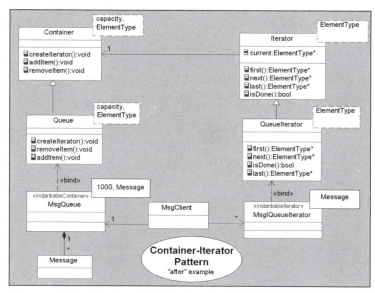

Figure 7.4 Details of use of container-iterator pattern

Interface Abstraction Pattern Strategy

The interface abstraction strategy separates the things that change from the things that don't. Specifically, the specification of services (operations and events) is separated from the implementation of the desired services. This allows the services to be provided in different flavors, each of which optimizes a different set of criteria, such as worst-case performance, average performance, space-complexity (i.e., memory usage), resource usage, etc.

The adaptor pattern is a prime example of the interface abstraction strategy. The situation arises when the client needs or expects a certain interface, but it is different from what is provided by the underlying realization. A simple example occurs when a track manager needs to get data from a variety of different sensors, but the sensors are from many different vendors and even nationalities. The solution is to require a common interface and write adaptors that convert service requests to the sensors and responses coming from the sensors to meet the expectations of the track manager. Figure 7.5 shows the structure of the adaptor pattern. You can see the client expects a certain interface (TargetInterface). The adaptor subclasses both the desired interface and the implementation (ActualInterface). Then the implementation of the TargetInterface services is specialized to implement the expected interface in terms of the actual implementation offered by the ActualInterface.

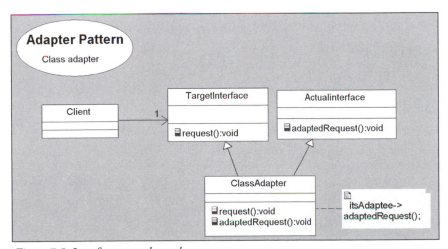

Figure 7.5 Interface example—adaptor pattern

Figure 7.6 provides another simple example of the use of the adaptor pattern. In this case, the client needs a stack, but what we have is a linked list. The IntStackAdaptor class subclasses both the IntStack and the Linked list. The inherited operations from IntStack are overridden to provide the desired functionality using the insert() and remove() operations of the linked list implementation.

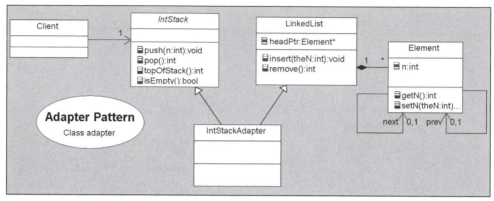

Figure 7.6 Use of adaptor pattern

Design patterns consist of four primary aspects:[7]

- Problem
  - What problem does the pattern address?
- Applicability
  - Which design criteria does the pattern optimize?
  - When is the pattern applicable (i.e., what is the common problem context)?
- Solution
  - The pattern specification
- Consequences of the pattern
  - Pros (benefits) of the pattern
  - Cons (costs) of the pattern

Mechanistic design patterns are medium scale, involving as few as two or as many as a dozen classes. The interested reader is referred to the references for more patterns.[8]

---

[7]   The exact number of pattern aspects differs depending on the author consulted. We believe these are the most important and general aspects.

[8]   See, for example, the Patterns Home Page at *http://hillside.net/patterns*. The aforementioned book by Gamma et. al. is highly recommended, although it is not cast in UML terms. There are mechanistic patterns in the author's books, such as *Doing Hard Time and Real-Time UML, Third Edition*. Another useful book is *Applying UML and Patterns* by Craig Larman, Prentice Hall, 1998. The author of the current book teaches a 3-day class on real-time design patterns. Interested parties should contain the author at *Bruce.Douglass@Telelogic.com*.

Detailed Design

Detailed design optimizes the system at the class level, and so is concerned primarily with class features—attributes, operations, states, activities, and ports. There are anywhere from 10 to 100 classes (and possibly more) involved in a collaboration but only a few of them really require special attention. This is because either they are more complex than their brethren, either in data structuring or in behavioral complexity, or because they are part of a control or data flow that has high optimization requirements.

Figure 7.7 shows the detailed design workflow as specified in the Harmony process. You can see that the classes are optimized one at a time. The specific optimization steps are between the fork and join, indicating that the order in which these optimizations are done is not specified (you get to choose ☺). Further, not all of these steps need

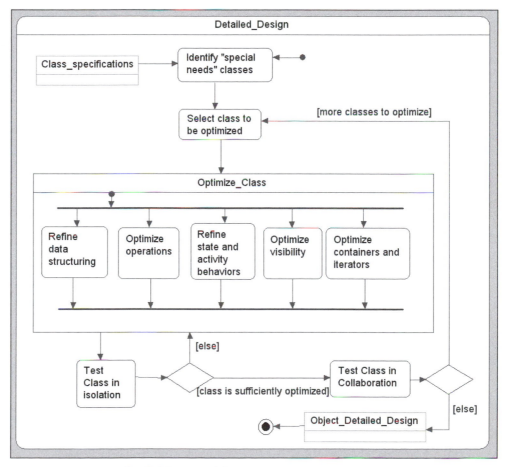

Figure 7.7 Harmony detailed design workflow

to be taken for each class under consideration. Some classes may require primarily data restructuring while others might require state machine optimization.

Like the preceding design phases, detailed design proceeds largely through the application of design patterns, although in the literature they are more commonly referred to as design "idioms." That is, for the most part, the problems you're trying to address in this most-detailed level of design have been solved before in different systems and perhaps for different applications. However, those solutions can be abstracted, generalized, and reapplied in new situations, including your current system under development.

The issues normally addressed during detailed design include:

- Ensuring operation pre- and post-conditional invariants, such as
  - Range checking
  - Reasonableness checking
  - Unit checking (e.g., miles vs. kilometers)
- Internal data structuring
- Internal algorithmic structuring and optimization
- Optimization of state machines
- Specifying local error and exception handling (both accepted and thrown)
- Low-level redundancy for safety and reliability
- Numeric round-off error management and correction
- Abstraction of services into interfaces
- Class feature visibility
- Ensuring quality of service budget adherence for services
- Realization of associations, e.g.,
  - Pointers
  - References
  - Object IDs
  - Widget handles
  - Sockets
  - …

The first five of these get most of the attention, but all are important. For example, consider the following general problem for a class state machine: the state machine wants to remember that an event happened in the past and handle it when it is convenient, or relevant, to do so. It is easy to use AND-states and see if another AND-state is in a particular state with the IS_IN(stateName)[9] operator, but how do you remember that an event or state was ever visited in the past?

One answer is the latch state pattern.[10] The structure of this pattern in shown in Figure 7.8. The state machine has three AND-states. The predicate AND-state is responsible for identifying the predicate event—that is, the event that must be remembered. The latch AND-state is responsible for remembering that the event occurred. The dependent AND-state depends upon the event that was remembered. When the predicate state receives an event that should be remembered, it generates a different event (the GEN() action in the predicate action list) that causes the latch AND-state to transition from the Unlatched to the Latched state. Later, when the dependent state receives an event that requires the original event to have been received at some point in the past, a guard checks to see if the latch is in the Latched state. If it is, then the guard is TRUE, and the transition can be taken. When the transition is taken, another event is generated that clears the latch by sending a clearLatch event, causing the latch to transition back to the Unlatched state.

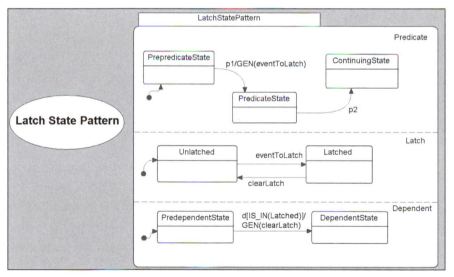

Figure 7.8 Latch state pattern

---

[9]   IS_IN(state-name) is a behavior of a reactive object, returning TRUE if the object is currently in the specified state and FALSE otherwise. Rhapsody supports this query.

[10]   The concept of state behavior patterns was introduced in a previous book by the author: *Doing Hard Time: Developing Real-Time Systems with UML, Objects, Frameworks, and Patterns*, Addison-Wesley, 1999.

Figure 7.9 shows a simple usage of the pattern. A sensor runs three AND-states. The top AND-state has the responsibility for acquiring the data when available. The bottom AND-state takes the data and performs some long computation on the data. Because this computation can take a long time, and because the bottom AND-state doesn't want to miss any incoming events when it can get around to processing it, the event is remembered by the latch AND-state in the middle.

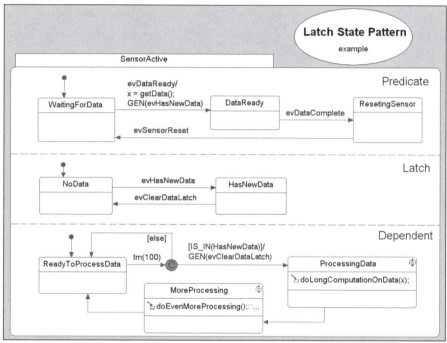

Figure 7.9 Use of latch state pattern

Other than the use of state machine behavioral patterns, most of the optimization done within detailed design is done "below" the UML modeling level. For example, range and validity checking is done by adding actions to methods, activity diagrams, and state machines.[11] There are a few other optimizations that manifest themselves in the model as well, such as the identification of interfaces, additional class-level

---

[11] The *action language* is the language used to specify primitive behavioral statements ("actions") in the UML. The UML specifies the kinds of things that an action language must be able to state in the UML action semantics specification, but does not specify or define a particular action language. It is most common is to use the implementation language (e.g., C, C++, Ada or Java) as the action language in the model. Some tools have proprietary action languages but these are not without problems, including difficulty in debugging and lack of reverse engineering code changes when necessary.

data redundancy, and feature visibility, but much of the optimization takes place by adding actions to achieve specific optimizations.

For example, suppose we have a class PatientInfo that has attributes of height, weight, age, sex, ID#, address, phone. Some of these are safety-critical (weight, age, sex, and ID), and so they can be stored redundantly. Further, additional checks can be made for reasonableness. Figure 7.10 shows how range and data corruption checks might be added in detailed design. The PatientInfo class has redundant one's complement storage of the safety-related values.[12] The comments show the actions within the mutators (setWeight()) and accessor (getWeight()) operations that perform range checking and check that the stored data was not corrupted.[13]

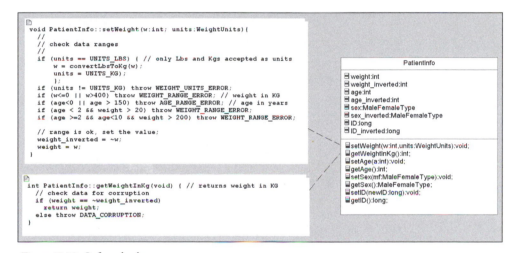

Figure 7.10 Safety checks

12   One's complement (bit-wise inverted) storage is used in case the problem is with stuck-at-0 or stuck-at-1 memory faults.

13   Corrupted data is a problem in electronically noisy environments. The author has personally witnessed memory bits flipping within a medical treatment device while in a medical operating room environment due to the operation of electrosurgical equipment.

Problem 7.1   Applying Mechanistic Design Patterns—Part 1

In this first problem, we'll apply certain mechanistic design patterns to the Roadrunner Traffic Light Control System and the CUAV. In the next problem, we'll take the solution further and add more design patterns to resolve more design issues and further optimize the solution.

There are many design patterns that can be applied to the systems we've been modeling—hundreds of possibilities, at least. In this problem, we'll discuss a couple of design problems and patterns that address it, and then you'll incorporate them in your model.

The Roadrunner Traffic Light Control System must interface with many different sensors from different manufacturers. Also, the system needs to be able to easily adopt new sensor modalities. To do this we need to support a general interface for both vehicle and pedestrian sensors and use adaptors to support different specific sensors and sensor technologies. The task for this part of the exercise is "Optimize the collaboration for support of different sensor modalities and technologies." So, for the first part of this exercise, apply the Adaptor pattern, discussed earlier in this chapter (see Figure 7.5 and Figure 7.6), to optimize the Detect Vehicle collaboration to support different sensors. Figure 7.11 shows the starting point, the object analysis model of the Detect Vehicle use case. Show the resulting mechanistic design-level class diagram.

For the second part of this problem, we'll address noise reduction and image enhancement in the image processing of the Reconnaissance Management Subsystem (RMS) of the CUAV. This is a crucial problem that must be solved for effective reconnaissance and track management. It may be necessary to add or delete different image-processing algorithms dynamically, based either on commands from the Mission Ops personnel or onboard by trying a best-fit or error minimization function. Some noise reduction and image enhancement algorithms are:

- Gaussian sharpening

  - Gaussian sharpening removes noise by convolving the original image with a mask. This brings the value of each pixel closer to that of its neighbors.

- Averaging Filter

  - Averaging is a degenerate case of Gaussian filtering, where the function defining the mask values has an infinite standard deviation.

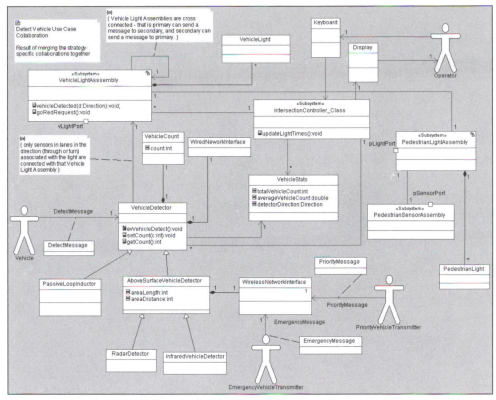

Figure 7.11 Roadrunner Detect Vehicle collaboration

- Median filter
  - A median filter is a nonlinear filter that can preserve image detail. A median filter works by:
    - considering each pixel in the image
    - sorting the neighboring pixels into order based upon their intensities
    - replacing the original value of the pixel with the median value from the list
- Center-surround neural network error minimization
  - A neural network is an ordered collection of rather simple processing units, arranged to perform error minimization. Neural networks usually require training (by applying examples with known results, hundreds or thousands of times). Neural networks come in several types. A center-surround neural network is the computational basis for image sharpening in the human visual system and works by sharpening edge boundaries in images.

There are, of course, many more image-enhancement algorithms from which to choose. The RMS may have a dozen or more available from which to choose. The RMS needs some way to efficiently switch algorithms or to chain algorithms together to provide optimal image processing. The goal for this part of the problem is to "Optimize the system for flexibility in supporting multiple image-processing algorithms in sequence."

To realize this goal, instantiate the chain of responsibility pattern, discussed in the sidebar.

Chain of Responsibility Pattern

Problem:

You want to have flexibility in the addition of new and possibly multiple algorithms ("handlers").

Applicability:

- When the best place to handle a request isn't known at design time
- When you want to support a multilevel response
- When you want to be able to dynamically modify the response
- When you want to be able to give a set of possible responses each a chance to handle the request

Solution:

Decouple the request handling from the receiver of the request by creating a chain of handlers and pass the request from handler to handler so that they each get a chance to respond until either the list of handlers has responded or all handlers have been given the chance to respond. Any specific handler can:

- Choose to handle the request and terminate
- Choose to handle the request and pass on the request to the next handler
- Choose not to handle the request and pass on the request to the next handler

Consequences:

- Reduced design-time coupling of the class that receives the request from the algorithms that process the request
- Ability to add request processing dynamically
- There is no guarantee that a request will be handled because there is no grand oversight

Structurally, the chain of responsibility is a very simple pattern, as we can see in Figure 7.12.

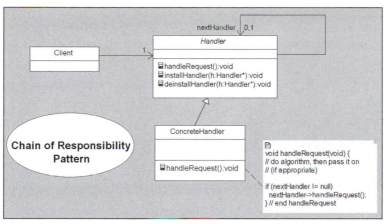

Figure 7.12 Chain of responsibility pattern

For this exercise, use the chain of responsibility pattern to improve the algorithmic flexibility of the image processing for the Reconnaissance Management subsystem, whose analysis class diagram is shown in Figure 7.13. Show the resulting mechanistic design-level class diagram. Hint: for this exercise, you need only show the elements from the collaboration that relate to the application of the design pattern.

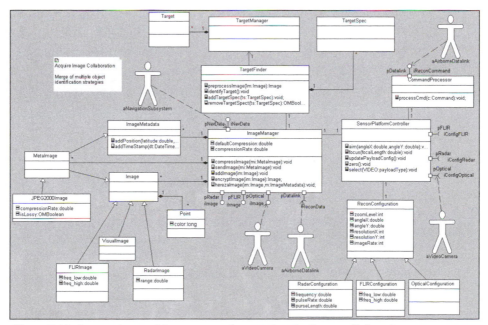

Figure 7.13 Reconnaissance Management subsystem Acquire Image collaboration

Problem 7.2   Applying Mechanistic Design Patterns—Part 2

In this next problem, we're going to continue applying mechanistic design patterns to the collaborations we've been working on. For the Roadrunner system, we want to optimize the timely update of change-of-state information of the lights to the IntersectionController, as well as to the other vehicle lights (from the safety architecture, it was determined that the lights need to share contextual state information to ensure an appropriate overall system state). It is true that the light assemblies can, whenever they change state, notify the IntersectionController and the other clients. However, we may want to dynamically add other clients of that information, even at run-time. In such cases, there must be a mechanism in place to let the light assemblies know to whom the state information should be sent.

A common solution to this problem is provided by the observer pattern (also known as "publish-subscribe pattern"). In this pattern, the servers (in this case, the light assemblies) don't know their clients a priori, but instead are instructed, at run-time, to add clients who wish the information. They do this by providing a subscription mechanism in which the clients subscribe and provide their contact information. When it is appropriate (in this case, when the light assembly state changes), the client list is walked and the information is sent to each subscriber. The observer pattern is one design solution that could be used in this case. However, we have a deeply cross-connected system. In the safety analysis done in the previous chapter, it was decided that each light will talk to the others and in this way ensure that the system overall will always remain in a safe state (that is traffic is never allowed to go in crossing directions at the same time).

Another option is the data bus pattern.[14] The data bus pattern optimizes the connection topology of deeply cross-connected architectures by providing an object that serves as a repository for shared data. In the "push" version, when the server changes state (or changes the value of interest), the data is pushed to the data bus, which in turn sends it to all the subscribers. This simplifies the construction and maintenance of the cross-connecting topology.

---

[14]   See the author's *Real-Time Design Patterns: Robust Scalable Architecture for Real-Time Systems,* Addison-Wesley, 2002. The pattern shown here is slightly modified from that reference.

Data Bus Pattern (Push Variant)

Problem:

Need to be able to share data among many different clients.

Solution:

This is a variant of the observer pattern, where a data bus object holds shared data and distributes the datum whenever a new value is "pushed" to the data bus. Clients subscribe to data of interest owned by the data bus.

Consequences:

- A simple pattern that allows the addition of clients at run-time and simplifies deeply cross-connected topologies.

- Additional memory is required to hold copies of the data and the notification handles (addresses) of the clients

Figure 7.14 shows the structure of the data bus pattern. The composition shows that the different composites own their own copies of the data. The pattern can be mixed with other patterns, such as proxy or broker, to enable the pattern in distributed architectures.

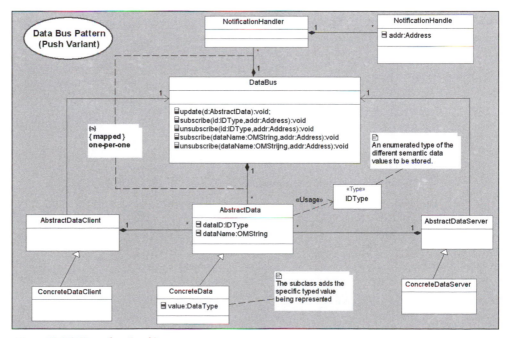

Figure 7.14 Data bus (push) pattern

Figure 7.15 shows a simple application of the pattern. In this case, two data elements are put into the data bus. The train speed sensor creates an instance of the VelocityData class and calls update() to update the value. The GPS object creates an instance of the PositionData class and similarly updates the copy owned by the data bus. The engine closed-loop control needs, and subscribes to, the velocity data while the display subscribes to both velocity and position. When the data is updated, the clients are updated automatically by the data bus. The DataBus:: update() operation locates the appropriate instance of which to update the value; if the ID or name can't be found, a new one is added to the composite along with a new notification handler.

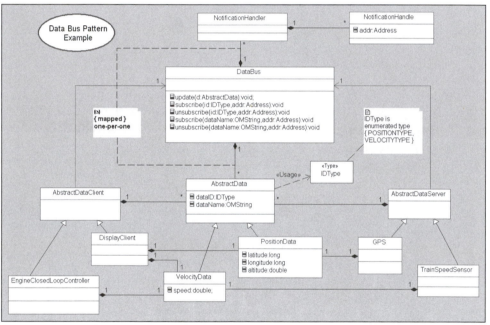

Figure 7.15 Data bus pattern example

Now to apply it in our problem … For this part of the exercise, apply the data bus pattern to manage the update of the states of the individual light assembly instances and distribute them to all the interested clients.

For the second part, let's consider again the CUAV Reconnaissance Management subsystem Acquire Image collaboration in Figure 7.13. We see that commands are sent from the Mission Ops actor to command the acquisition process in a variety of ways:

- Configure sensor

- Orient sensor

- Request image

- Configure image processing

- Set target types and their properties

These commands must persist until handled. The command pattern provides a design means to address this kind of object interaction. The pattern is described in the sidebar. It works basically by reifying the command or requesting itself as an object, allowing it to be remembered and manipulated. Figure 7.16 shows the basic structure of the command pattern, while Figure 7.17 shows a simple example.

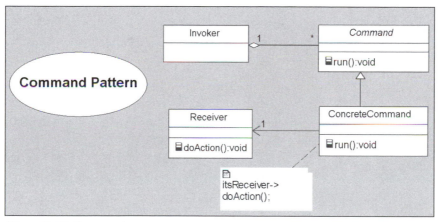

Figure 7.16 Command pattern

Command Pattern

Problem:

You want to manage requests that require logging, queuing, or take a period of time to fulfill.

Applicability:

- When you want to take a request through a series of states (steps)
- When you want to manage multiple requests simultaneously, each in possibly different states
- When you want to support undo
- When you want to log or queue requests

Solution:

- Reify the request as an object, possible with internal state
- Aka "transaction pattern"

Consequences:

- Allows simultaneous management of multiple requests easily
- Commands are first-class objects and can be subclassed and extended
- "High-level" commands may be assembled from smaller command parts
- Easy to add new commands

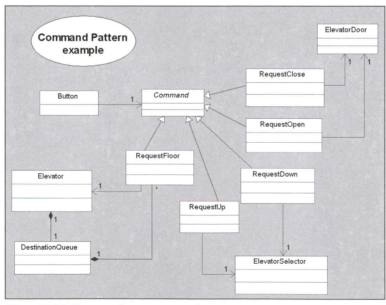

Figure 7.17 Command pattern example

For this latter part of this problem, apply the command pattern to the CUAV to specify and optimize how it handles and manipulates commands.

Problem 7.3 Applying Detailed-Design State Behavior Patterns

The next problem brings us into the detailed-design phase. As discussed early in the chapter, detailed design is all about optimization at the class or object level of scope. This problem will deal with the optimization of class state machines to solve particular detailed-design issues. That is, we will look for opportunities to apply state design patterns to the state machines in the classes of the collaborations to optimize or simplify their structure and/or behavior.

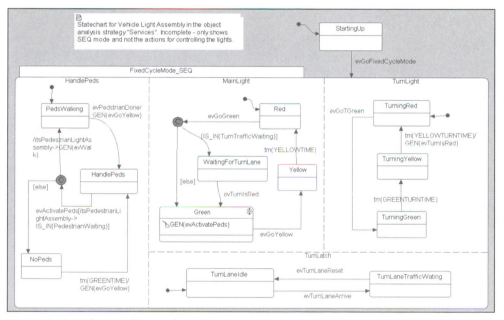

Figure 7.18 Light assembly statechart

Figure 7.18 shows the statechart we presented previously for the vehicle light assembly. We note that this state machine already uses the latch pattern discussed earlier in this chapter. We know that the light assemblies must support more modes than this, including (see the use case chapter or Appendix A for more detail):

• Safe intersection mode (all lights flash red)

• Evening low volume mode (primary flashes yellow, secondary flashes red)

- Fixed cycle time mode (fixed time cycling of lights)

- Responsive cycle mode (lights cycle according to traffic)

- Adaptive mode (like fixed cycle time mode except that the length of the cycle times vary depending on traffic load)

In general, whenever you see the word "mode," think "state." This means that, at the high level, the lights must support several modes of operation:

- Off

- Safe (Flashing Red)

- Cautionary[15] (Flashing Yellow)

- Fixed Cycle

- Responsive Cycle

- Adaptive time Cycle

Moreover, we must be able to transition directly from any of these high-level states to any other. If we draw such a state machine, it looks like Figure 7.19. The term "rat's nest" comes to mind.[16] Isn't there a cleaner, simpler way to represent exactly the same behavior?

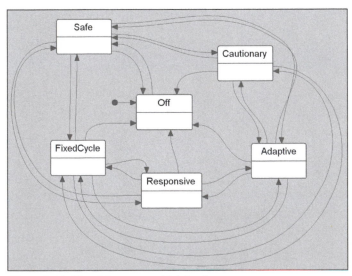

Figure 7.19 Transition among high level states

---

15   Note that these states are from the light assembly's point of view. In Evening Low Volume (system) mode, the primary light would be in Cautionary state while the secondary light would be in Safe state.

16   I suppose rats have to have a place to live, too—I just don't want them in my software!

The any state pattern uses composite states to simplify the transition structure while providing exactly the same behavior. It accomplishes this by placing the semantic states inside a composite and drawing a single transition from the composite state to the internal semantic state. That transition represents all transitions from the peer semantic states because whenever the object is "in" the composite state, the transition may be invoked, regardless of exactly which nested semantic state is active. Figure 7.20 shows the rat's nest state machine before, and the streamlined and exactly equivalent state machine after encapsulating the semantic states with the composite.

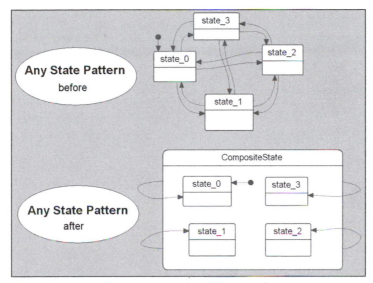

Figure 7.20 Transitions among high-level states with any state pattern

For this part of the exercise, apply the any state pattern to the light assembly state machine; be sure to include the appropriate events, guards, and actions (as necessary) on the state machine. The high-level states don't need to be detailed on the same state machine; in fact, I recommend the use of submachines (state machines drawn on a different diagram but represented as a state on the high-level diagram) to make the diagrams more readable.

For the Coyote UAV, we haven't built any state machines yet for the Acquire Image collaboration (shown in Figure 7.13). In this exercise, you will construct a statechart for the ImageManager to acquire and process image data from the radar, FLIR, and optical payloads. To solve this problem, do the following:

- Treat the ImageManager as a composite class and create a structure diagram for this class.

- Add in three parts to manage the acquisition and image processing of the payload image data:
  - theRadarDeviceDriver:RadarDeviceDriver
  - theFLIRDeviceDriver:FLIRDeviceDriver
  - theOpticalDeviceDriver:OpticalDeviceDriver
- In a separate class diagram, create a class called SensorPayloadDeviceDriver as the base class for these three classes and make the specific device driver classes subclasses of this.
- In the SensorPayloadDeviceDriver class, add a statechart that has the following states:
  - Off
  - WaitingToGetRawImage
  - ConstructingImage
  - PreprocessingImage
  - AddingMetadataState
- Link the states with the following transitions:
  - In this state machine, when it is off an evEnableSensor event takes it to the WaitingToGetRawImage state.
  - From there, evImageStart event takes the machine to the ConstructingImage state, executing action "initializeImage()".
  - From that state, an evImageData event transitions back to the ConstructingImage state, executing the action "addDataToImage()".
  - From that state, an evImageComplete event transitions to the PreprocessingImage invoking the handlers added when we added the chain of responsibility pattern earlier in the chapter by executing the action imageHandler->handle().
  - When that processing is complete, transition to the AddingMetaDataState and invoke the action "theImage->addMetaData().
  - Once that is complete, cycle back to the WaitingToGetRawImage state.
  - If, at any time (except in the Off state), new data becomes available (i.e., an evImageStart event is received), abort the current image and cycle back to the ConstructingImage state and execute the action "initializeImage()".
  - If, at any time (except in the Off state), an evDisableSensor event is received, abort whatever image processing is proceeding and transition to the Off state.

This is the first half of the problem. Now we have to address the concern of how we get the data. Let's suppose that we want to get the data by checking periodically, a process known as polling. How do we effectively poll on a periodic basis?

One solution is the polling state pattern. This pattern consists of two and-states. The first, called the "DataHandlingRegion" in Figure 7.21, is where the semantic handling of the data occurs. When it commanded into the Active state, it sends an event, which is consumed by the second and-state. This latter and-state focuses on its responsibility to periodically check for new data. When the POLLTIME period has elapsed, it invokes the acquireData() action, polling for the data, and then generates the evDataReady event. The first and-state consumes that event and transitions to the CrunchingData state to handle it.

The polling pattern simplifies the overall behavior by segregating the semantic data manipulation from the work of acquiring the data periodically.

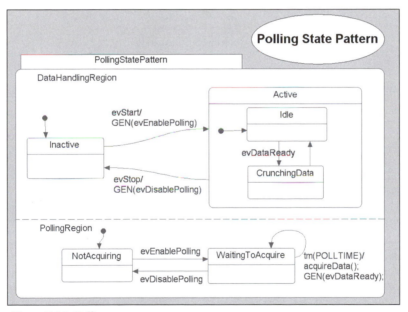

Figure 7.21 Polling state pattern

Figure 7.22 shows a simple application of the polling state pattern for acquiring air speed data. When it starts, it powers the hardware and enables polling. When data is acquired from the air speed sensor, the second and-state generates an event that causes the semantic region to filter and then queue the manipulated data.

Let's now apply this pattern in the CUAV model. For the latter part of the exercise, add the polling pattern to the statechart you've constructed to add polling behavior. Just to make it interesting, the acquireData() operation returns a value called success which can be one of three values: 0 (NONE), in which case data is not ready; 1 (NEW_IMAGE) in which a new image is ready to begin (i.e., generate a evImageStart event), or 2 (ADD_DATA) in which case the next part of the current image is ready (i.e., generate an evImageData event). Incorporate this functionality into the resulting state machine.

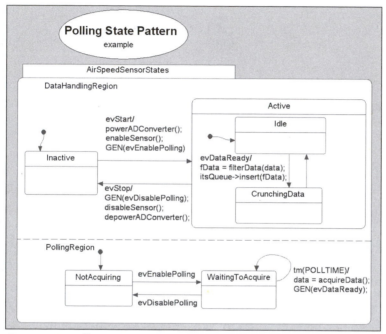

Figure 7.22 Polling state pattern example

Problem 7.4   Applying Detailed Design Idioms

For this last problem, I'd like to address some detailed design idioms. There are many small detailed optimizations that are typically performed in detailed design. These are listed earlier in the chapter. In this last problem, we will deal with two detailed-design issues. First, how does one handle errors or exceptions in the UML, especially in state machines, and second, how do we ensure adherence to preconditional invariants on service requests?

The common idiom in C to handle errors is to return a nonzero value from a function. The problem with this idiom is that every client is free to ignore those error indications, and I've debugged more than my fair share of defects in safety-critical embedded systems in which error indications were simply ignored by the clients. Exception handling in C++ and Java, though, forces the client to handle the thrown exception or the calling function terminates. Once an exception is thrown from a server, the application looks for an appropriate exception handler (i.e., "catch clause") in the caller; if one is found, then the exception is handled. If not, then the caller terminates and the application looks for an exception handler in its caller, ad infinitum. This process is commonly called "walking back the stack." Eventually, either an exception handler is found, or the application terminates.

Such language-specific exception-handling mechanisms have two things going for them. First, they cannot be ignored and that is crucial for high-reliability and safety-critical systems. Secondly, they separate the exception handling from the main functionality, simplifying both. When the two are intermixed, the logic for both the main processing and the exception handling is muddied and hard to understand. By separating out the two, both become simpler and easier to get right.

However, these exception-handling mechanisms are not without their problems. The most significant of these is their inability to handle exceptions across thread boundaries. In C++, if an exception is not caught by the time the stack has unwound to the top of the thread, then the thread must terminate, and then the application must terminate. Another problem is that it isn't entirely clear how to use exceptions in the presence of state machines.

State machines in UML may be synchronous (using call events, or what Rhapsody calls "triggered operations"), asynchronous (using signal events, or what Rhapsody calls "receptions"), or a combination of the two. If a series of call events affects multiple state machines and an exception is thrown, then the stack gets walked back just like it would with normal method calls. It's a little more complex because care must be taken to ensure each state machine remains in a valid state, but it is relatively straightforward. It is much less clear what happens if a signal event occurs. In that case, the stack walks backwards through the active object owning the event queue, but it would be odd indeed to have the active object understand the appropriate exception-handling behavior for all the objects held within that thread boundary. So what to do?

In the UML, an exception is considered to be a kind of signal. In the UML 2.0 Final Adopted Specification, it defines an exception as:

> *A special kind of signal, typically used to signal fault situations. The sender of the exception aborts execution and execution resumes with the receiver of the exception, which may be the sender itself. The receiver of an exception is determined implicitly by the interaction sequence during execution; it is not explicitly specified.*

Even in this case, the receiver of the exception is implicit—that is, it is the service requester. In general, an asynchronous state machine doesn't know who its sender is, unless that information is provided as a parameter for the passed signal event.[17]

So how should we handle exceptions when we have multiple threads and when we have state machines? Whatever mechanism we use, we would very much like it not to depend on whether the source event is synchronous or asynchronous, because having multiple ways to manage exceptions complicates the applications and makes it pathologically tightly coupled. And, while we're at it, we need to be able to handle normal language exceptions in our state machines as well, because actions may invoke services—some of which are in the standard libraries, such as new()—that may throw exceptions.

One solution is the exception state pattern. This pattern uses events to indicate not only service invocations but also exceptions. The class diagram in Figure 7.23 shows structural elements involved. The StatefulClient and StatefulServer are both reactive classes; that is, they use state machines to specify their behavior. They also connect with ports. This is important because the StatefulServer will use the port

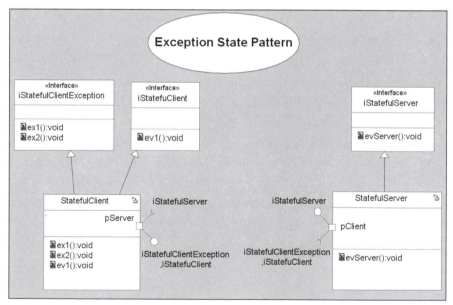

Figure 7.23 Exception state pattern

---

17   An exception (pardon the pun) to that rule is that when ports are used, the source port for an event can be identified, but even then, it is the port that is identified, not the actual event source. Nevertheless, if the port is bidirectional, the port can be navigated backwards, provided that the exception is passed as a signal event that is specified in the interface contract of the port.

name to identify along which link the exception event should be sent. The same event could be from different sources, but as long as each connects via a different port, then the source can be identified.

The StatefulClient realizes two distinct interfaces. The first, iStatefulClient, is the semantic interface. It specifies the events normally received during the execution of the object's life. The second, iStatefulClientException, is the exception interface—these are the events that have to do with the handling of faults and unexpected conditions and occurrences. The StatefulServer is typed by a single interface, iStatefulServer.

The ports connecting the instances of the classes are typed by these interfaces. The StatefulClient port pServer requires the iServer interface and offers both the iStatefuleClient and iStatefulClientException interfaces. Using two distinct interfaces helps segregate the normal semantics from the exception handling. The StatefulServer port pClient requires both these interfaces, while offering its own semantic interface, iStatefulServer.

The state machine for the StatefulServer is provided by Figure 7.24. We see how the exception gets raised in the action list for the evServer event. If the doAction returns an unsuccessful result, it generates an ex1 event[18] and sends it back to the original sender. The original sender is identified by the guard on the event; IS_PORT(pClient) identifies that this event is from the pClient port. If the event can come from other ports, then they must be listed in other transitions. Following the sending of the exception event back to the client, an evRollback event is generated to return to the original state.

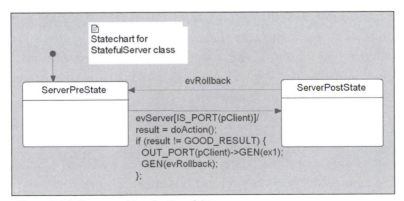

Figure 7.24 State machine for StatefulServer

---

[18]   We typically use the ev prefix for normal semantic events, and the ex prefix for exception events.

Lastly, the state machine for the StatefulClient is shown in Figure 7.25. We notice that the semantic actions are segregated from the exception actions by putting each in their own and-state. When the ev1 event is handled, it sends an evServer event out the pServer port. If it later receives an exception event (ex1 or ex2), then it does whatever clean-up is required and then issues an evRollback event to return to the PredecessorState.

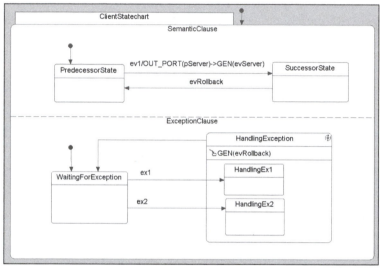

Figure 7.25 State machine for StatefulClient

We see that the pattern works by using a separate set of events, collected into the iStatefulClientExceptions interface, to represent the exceptions being thrown. The server uses the IS_PORT() in a guard to identify the source of the event that led to the exception identification. We create a separate and-state to handle the exception events. And in both the client and the server, we typically want to do a rollback to the previous state in the case of an exception being identified. The exact exception-handling behavior is, of course, case- and situation-dependent but such specific handling behavior can easily be handled.

In the Roadrunner Traffic Light Control System, the light assemblies associate with the actual lights themselves. This association is a composition, indicating strong ownership. In this special case, there is a navigable link between the light and the composite light assembly. Since no other object can directly command the lights, ports aren't required. All that is necessary is that, if the VehicleLight or PedestrianLight has an exception, it passes the appropriate event back to its composite owner.

The relation between the VehicleLightAssembly and the VehicleLight has *
multiplicity, one for each traffic lane in the "through" direction and another for the
turn lane. The PedestrianLightAssembly has a single light that it controls, so the
multiplicity needs to be changed to 1. These changes are reflected in Figure 7.26.

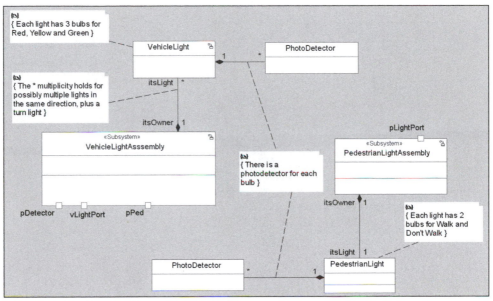

Figure 7.26 Light assemblies and lights

For this exercise, add the statecharts for the vehicle and pedestrian lights (the
lights are dumb, so any state may be entered from any other state—sequencing is
handled by the light assemblies). Then add the exception state pattern to handle
the error condition that when a bulb is illuminated, the photo detector does not
detect the light going on. As mentioned, ports aren't required because the light
assembly has a composition relation with the lights, and that composition provides
a navigable link at run-time.

Next, let's work on the CUAV. Previously, we added the polling state pattern to
the Acquire Image collaboration for the Reconnaissance Management Subsystem for
the CUAV. In this exercise, we want to avoid a problem commonly known as "lock-
up"—that is, the condition in which a system stops what it is doing, either because
of a software or hardware fault. The solution we will add into the collaboration is
called the watchdog pattern. The watchdog pattern is detailed in the sidebar.

Watchdog Pattern

Problem:

Need to ensure liveness of periodic processes.

Solution:

Run an observer in a separate thread that receives "life ticks" on a periodic basis. If the watchdog doesn't receive the life tick in a timely fashion, the system is assumed to have locked up and some draconian response, such as system (or subsystem) reset, or transition to fail-safe state, is taken.

Consequences:

* Useful only when the timeliness of the monitored processes can be predicted with fair precision.

* Different watchdogs can be used for different processes when specialized responses are required to different process faults.

* Many different processes may be monitored.

The watchdog pattern works by having a watchdog active object (or it runs in its own thread) monitor clients. The client must subscribe to the monitoring services by invoking the addMonitor(timeout:long) operation. This operation results in the creation of a monitor (called a WatchPuppy in the pattern) with the appropriate timeout and returns the ID of the monitor. This ID is used to identify which client is stroking which watchdog via the stroke(ID) operation. This results in an event being generated to the appropriate WatchPuppy. Should a given WatchPuppy timeout, it invokes the timeoutFault(ID:int) operation. The watchdog then in turn invokes the handleFault() operation on the SafetyExecutive, which takes whatever corrective action is appropriate. In the simple version, any fault results in the same action, e.g., a subsystem reset, but since the WatchPuppy ID is returned, more elaborate client-specific behavior could be easily added. Note that care must be taken that the timeout should be not less than the period of the process plus whatever jitter[19] occurs in that period.

Also note the use of the WatchPuppy puppyID attribute as a qualifier. This means that the puppyID is used to discriminate among multiple WatchPuppy instances;

---

[19] Jitter is the variance in the period of a periodic process. It is normally specified as an absolute value. A periodic process might have a period of 30ms +/− 2ms, where 30ms is the period and 2ms is the jitter.

the most common implementation is to make an array of WatchPuppies and the puppyID is simply the index into that array.

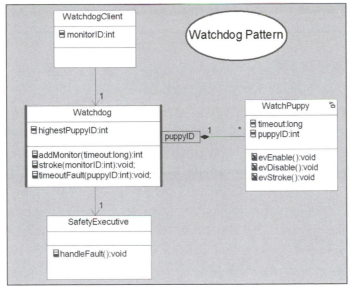

Figure 7.27 Watchdog pattern

The state behavior for the WatchPuppy is straightforward. As long as it receives the evStroke event frequently enough, the event arrives before the timeout fires. When the transition handling the evStroke event is processed, the state is exited and then re-entered, causing the timeout to be reset. The timeout fires only when the WatchPuppy hasn't received an evStroke event soon enough. This state machine is shown in Figure 7.28.

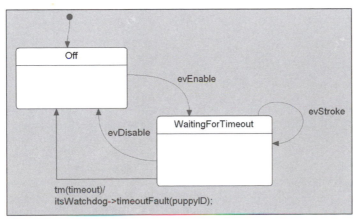

Figure 7.28 WatchPuppy state machine

For this last exercise, add the watchdog pattern to the Acquire Image collaboration to detect if the sensor device drivers fail for some reason. When this occurs, the SafetyExecutive should destroy and restart the ImageProcessing object. Show the resulting class diagram, and the updated state machine(s).

Summary

In this chapter, we have explored both mechanistic and detailed design. As with architectural design, these processes are about optimizing the system against various design criteria. Mechanistic design is focused at the level of the mechanism or collaboration, while detailed design focuses on the individual object. There is, naturally enough, some blurring of the lines among architectural, mechanistic, and detailed design. Such is the nature of artificial taxonomies, so you will occasionally come across a design decision that is the software equivalent of the platypus and doesn't fit neatly into the normal taxonomy. However, the taxonomy is useful because the literature is itself implicitly organized that way, and the use of the taxonomy makes finding the relevant patterns easier.

All of design is concerned with optimization and most of the design approaches are not fundamentally innovative. They are adaptations of previously used solutions. When a solution can be abstracted and reused in a different context, it becomes a pattern. The notion of patterns is so helpful that there are dozens of books and thousands[20] of web pages devoted to them.

This chapter wraps up this workshop. We've walked through the Harmony process from identification of use cases all the way through detailed design. I hope that it's been an interesting, useful, and enjoyable trip! ☺

---

[20]   A quick Google of "design patterns" returns about 14,000,000 hits!

Specifying Requirements: Answers

Answer 3.1   Identifying Kinds of Requirements

This exercise should hold no difficulties. It is a simple matter to create the requirements elements and link them together in a taxonomy with stereotyped dependencies. In this case, I split the requirements into two diagrams, as shown in Figure 8.1 and Figure 8.2. Note that I used the Rhapsody feature to insert a bitmap image showing the arrangement of the system elements in context.

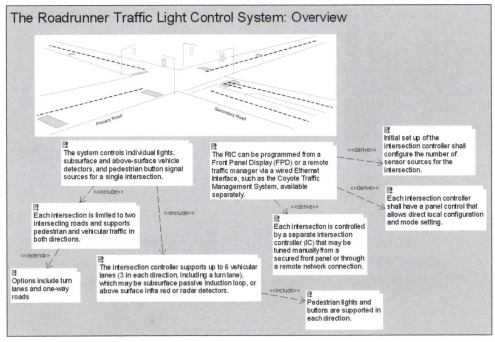

Figure 8.1 Overview requirements diagram

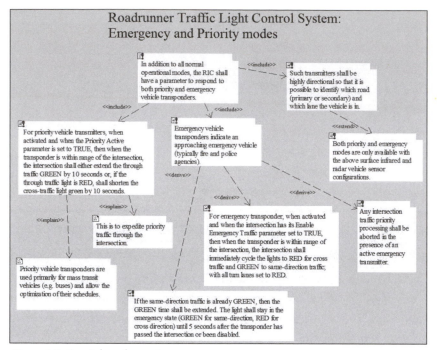

Roadrunner Traffic Light Control System:
Emergency and Priority modes

In addition to all normal operational modes, the RIC shall have a parameter to respond to both priority and emergency vehicle transponders.

Such transmitters shall be highly directional so that it is possible to identify which road (primary or secondary) and which lane the vehicle is in.

For priority vehicle transmitters, when activated and when the Priority Active parameter is set to TRUE, then when the transponder is within range of the intersection, the intersection shall either extend the through traffic GREEN by 10 seconds or, if the through traffic light is RED, shall shorten the cross-traffic light green by 10 seconds.

Emergency vehicle transponders indicate an approaching emergency vehicle (typically fire and police agencies).

Both priority and emergency modes are only available with the above surface infrared and radar vehicle sensor configurations.

This is to expedite priority traffic through the intersection.

For emergency transponder, when activated and when the intersection has its Enable Emergency Traffic parameter set to TRUE, then when the transponder is within range of the intersection, the intersection shall immediately cycle the lights to RED for cross traffic and GREEN to same-direction traffic; with all turn lanes set to RED.

Any intersection traffic priority processing shall be aborted in the presence of an active emergency transmitter.

Priority vehicle transponders are used primarily for mass transit vehicles (e.g. buses) and allow the optimization of their schedules.

If the same-direction traffic is already GREEN, then the GREEN time shall be extended. The light shall stay in the emergency state (GREEN for same-direction, RED for cross direction) until 5 seconds after the transponder has passed the intersection or been disabled.

Figure 8.2 Special mode requirements

Answer 3.2 Identifying Use Cases for Roadrunner Traffic Light Control System

Figure 8.3 shows a set of use cases for the Roadrunner Traffic Light Control System. The two primary use cases are Configure System and Manage Traffic. The Configure System use case interacts with either the operator of the front panel or the remote monitor. Independently of how the configuration takes place, the system also, of course, manages traffic.

The relations among the use cases were arranged very deliberately. While it would have been possible to just create a single Manage Traffic use case with a state machine providing the different modes as high-level composite states for that use case, that state machine would have been highly complex. I felt it was better to create subclasses for the different kinds of traffic management modes and have a simpler state machine for each. In this case, the intersection can switch among these modes but as far as the operational usage of the system is concerned, they represent separate

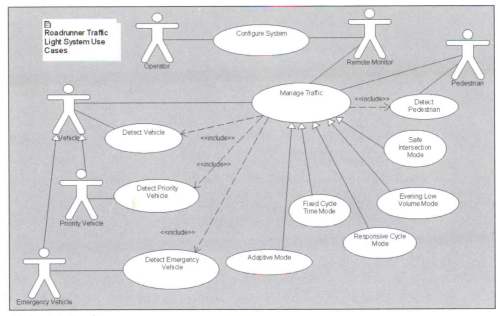

Figure 8.3 Roadrunner use cases

independent uses of the system, hence different use cases. Note that each of these use cases meets the criteria we set up in Chapter 3:

- It returns a value to at least one actor[1]
- It contains at least three scenarios, each scenario consisting of multiple actor-system messages
- It logically contains many operational requirements
- It does not reveal or imply anything about internal structure of the system
- It is independent of other use cases and may (or may not) be concurrent with them

Of course, most of these traffic-management modes require the ability to detect vehicles and pedestrians. These are modeled as use cases as well, but since they both contribute several of the Manage Traffic use cases, they are clearly at a lower level of abstraction than, say "Evening Low Volume Mode." We use the «include» stereotype of dependency to show this relationship. We could possibly have made Detect Emergency Vehicle and Detect Priority Vehicle more specialized use cases of Detect Vehicle, but in fact the operational effects of emergency and priority

---

[1]   Note that the specialized use cases inherit the base use case relations to the actors.

vehicles are fundamentally different and are likely to be used in different ways from an operational perspective. So while an argument can be made for modeling the use cases that way, I chose not to. The point is to model the requirements organization in a way that is consistent, makes sense, and can be used to move forward into object analysis and design. There are a number of good organizations that can be used for that purpose.

Notice also the specialization of the actor Vehicle. Emergency and priority vehicles are special kinds of vehicles that have special associations with particular use cases as well as the more generic relation due to the fact that they are, in fact, vehicles. While it is fair and reasonable to show generalization/specialization relations among actors, we do not show associations between them. Why? First of all, they are outside our scope and we have enough work to do just specifying and designing the parts we need to construct. Secondly, since it is outside our scope, we have little, if any, control over those interactions. If they should change, then our model will be wrong. Better to focus on the parts we need to build.

What about the choice of the actors? Why these actors? Why not "button," "traffic light," "walk light" and "vehicle sensor" instead? The rule of thumb I use to select actors I called the "In the Box" rule. If the object is inside the box that I ship to the customer—regardless of whether it is electronic, mechanical, or software—then it is an object internal to the system. If it is something that either the customer provides or is found at the operating site, then it is an actor. An actor, then, is the object that interacts with the things found in the box but is itself not in the box. The problem statement defines the lights, buttons, and sensors so I'm assuming that they will be in the box shipped to the customer to be installed at the intersection. This is always a crucial question to answer, because as developers and manufacturers we need to clearly understand the system boundary and what we are designing. If the problem statement said "interface with Acme Traffic Light Models XXX and YYY" then those would be the actors and not the vehicles. Having said that, an actor is any object—vegetable, animal (e.g., human), mineral (e.g., silicon-based hardware device) that interacts with the system in ways that we care about.

One last point about the actors—what about the Remote Monitor actor? The specification says that the system can be commanded via a network. Note that this actor isn't a network interface card (NIC), but the element that uses the NIC to do something interesting. We want to capture the proper actor in terms of the semantics of the interaction and not the technology used to communicate; the technology we will deal with in our design.

Additional Questions

- What is the problem with having the two separate use cases "Manage Pedestrian Traffic" and "Manage Vehicular Traffic?"

Answer 3.3 Mapping Requirements to Use Cases

For this problem, I selected the Detect Vehicles use case. I cut and pasted each individual requirement from the problem statement into separate requirements elements and linked them to the use case with dependencies. I felt that it was useful to decompose the use case into two sub-use cases, one for subsurface passive loop detection and one for above-surface detection. I also pasted in the bitmap to show the ranging of the infrared and radar detectors because I felt it augmented understanding. Note that it is common to use requirements traceability tools, such as DOORS™ from Telelogic to represent the dependencies between requirements and other model elements as "traceability links." Rhapsody can export and import these links to and from DOORS.

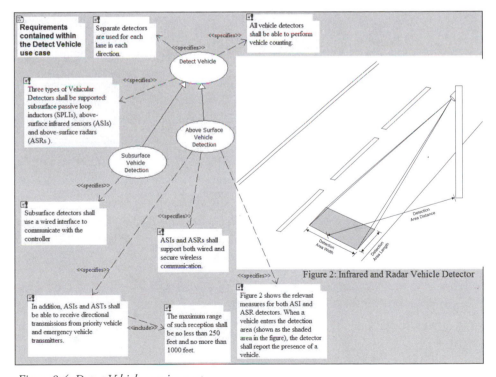

Figure 8.4 Detect Vehicle requirements

Answer 3.4 Identifying Use Cases for Coyote UAV System

The CUAVS is a large system. If we had all the use cases identified at the same level of abstraction, we might end up with, by the end, hundreds of use cases. Since the purpose of use cases is to organize the operational requirements into usable, coherent chunks, this doesn't meet the need. In the figures below, we have identified the "high-level" use cases and then decomposed them on separate diagrams. We also added a package to hold the parametric requirements.

Figure 8.5 shows the highest-level use cases. All but one of these (the Manage Datalink use case) will be decomposed on subsequent diagrams. In Rhapsody, you can add hyperlinks as user-defined navigational links to link the use cases with the more detailed use-case diagrams. Also note the package SystemParametrics; it contains a number of nested packages, each of which contains requirements elements and diagrams relevant to their subject matter.

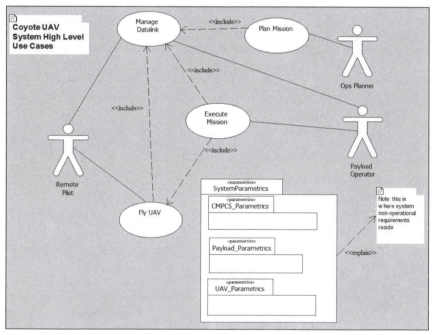

Figure 8.5 Coyote UAV system high-level use cases

Figure 8.6 shows the decomposition of the Execute Mission use case. This use case contains (via the «include» relation) the Perform Surveillance and Process Surveillance Data use cases. The use case is specialized into two forms—Execute Reconnaissance Missions and Attack missions. Each of these use cases is further decomposed. The former is specialized into Execute ECM (electronic counter measures), Execute Preplanned Reconnaissance, and Execute Remote Controlled Reconnaissance. Preplanned reconnaissance means uploading and following a flight plan which might search an area, along a route, or orbit a point target (such as a fixed defensive site).

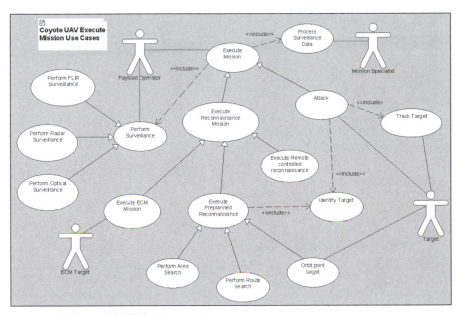

Figure 8.6 Coyote UAV Mission use cases

The Fly UAV use case is decomposed in Figure 8.7. It is important to remember that each of these leaf-level use cases is still a system-level use case. That means that each contains many specific requirements, many scenarios, and is probably detailed with a state machine or activity diagram.

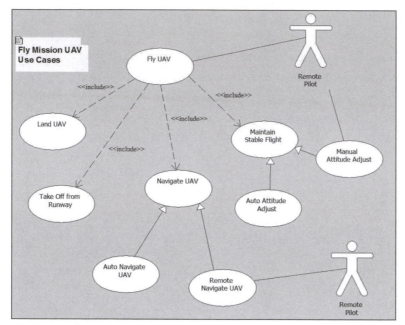

Figure 8.7 Coyote UAV Fly UAV use cases

Additional Questions

- What would the use-case model look like if it was drawn as a flat organization? Draw the diagram of that model and compare and contrast the presented hierarchical model with it. Which is easier to use and navigate?

Answer 3.5   Identifying Parametric Requirements

This exercise is very straightforward and is meant to provide practice in the organizing of such parametric requirements, as well as their representation in the model. In software-only projects, such requirements are frequently ignored, but in systems projects, they must be represented and tracked. Note that although this is created as a "use case diagram," it actually contains no use cases!

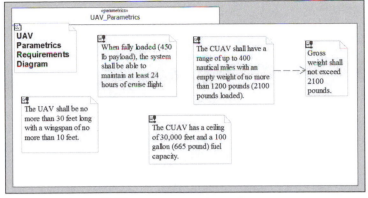

Figure 8.8 Coyote UAV system parametric requirements

Answer 3.6   Capturing Quality of Service Requirements

QoS requirements are a kind of constraint that is applied to another requirement or to a use case. In real-time systems, QoS requirements are crucial because they represent quantitative requirements such as performance, throughput, and bandwidth—issues vital to any real-time system. In this system, the problem statement identifies a number of QoS requirements around the operation of the vehicle, such as maximum altitude, maximum speed, cruise speed, and so on. If the requirement is not operational, then it should be placed in the UAV_Parametrics package but if they are operational then they should relate to the appropriate use case. In this case, this is for the Fly UAV use case. I created a use-case diagram and put the previously defined Fly UAV use case on it and added the QoS requirements to it. They relate to the use case using the «qualifies» stereotype of dependency. This is shown in Figure 8.9.

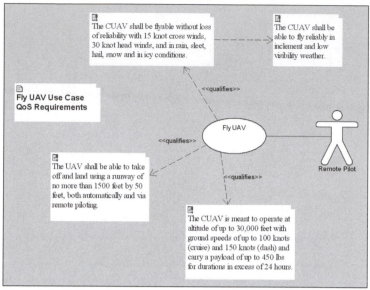

The CUAV shall be flyable without loss of reliability with 15 knot cross winds, 30 knot head winds, and in rain, sleet, hail, snow and in icy conditions.

The CUAV shall be able to fly reliably in inclement and low visibility weather.

Fly UAV Use Case QoS Requirements

<<qualifies>>

Fly UAV

Remote Pilot

<<qualifies>>

The UAV shall be able to take off and land using a runway of no more than 1500 feet by 50 feet, both automatically and via remote piloting.

<<qualifies>>

The CUAV is meant to operate at altitude of up to 30,000 feet with ground speeds of up to 100 knots (cruise) and 150 knots (dash) and carry a payload of up to 450 lbs for durations in excess of 24 hours.

Figure 8.9 QoS requirements

Answer 3.7   Operational View: Identifying Traffic Light Scenarios

The use case selected for this exercise was the Response Cycle Mode use case. In this use case, the system runs a fixed cycle mode except when a vehicle or pedestrian signals to use the intersection. Many different scenarios suggest themselves, such as:

- No traffic (fixed cycles only)

- Vehicle approaches from road B when road A has green

- Vehicle approaches from road B when road A has green and primary pedestrian light is WALK

- Vehicle approaches from road B when road A has green for its turn light

- Vehicle approaches from road B when road A has green for its turn light and a pedestrian is waiting to walk parallel with road A

- Pedestrian approaches from road B when road A has green

- Pedestrian approaches from road A when road A has green

- Vehicle approaches from road A to turn when road A has green

- Vehicle approaches from road B when road A has green; once road A turns yellow, a pedestrian approaches on road A

- Vehicle approaches from road B when road A has green; once road A turns yellow, a vehicle approach on road A

Then, of course, there's dealing with emergency and priority vehicles, system faults, and so on. One can easily think of a few dozen scenarios that show different actor-system interactions entirely contained within this use case.

To answer the problem posed, we only need to construct three scenarios. In our answer, we will elucidate the following scenarios:

- No traffic (fixed cycles only)

- Vehicle approaches from road B when road A has green and primary pedestrian light is WALK

- Pedestrian approaches from road B when road A has green

There are some interesting things even in the simple, no-traffic scenario, shown in Figure 8.10. First, note that the "system" is represented with the use case as a lifeline. This is perfectly reasonable and it can be interpreted as "the system executing the use case." Some people find that putting a use case here is counterintuitive, and so they prefer to use the «system» object as the lifeline. This is done in the third scenario; however, the meaning is equivalent. In either case, the lifeline stands for the system executing the use case.

Next, note that the actors in the scenario match the actors from the use-case diagram, shown in Figure 8.3. Because the Responsive Cycle Mode is a specialized form of the Manage Traffic use case, it inherits the actors used by the latter. To create the empty scenario, the use case and the actors were dragged from the Rhapsody browser onto a blank scenario diagram and then the details of the scenario were added. Because the scenario is an exemplar, it refers to specific actor instances. There are two vehicle actors, one for the primary direction and one for the secondary direction. Similarly, there are two pedestrian actors. In this first simple scenario, the system just provides autonomous behavior based on time, so we don't really need the actors to be shown on the diagram. However, we will need them for the other scenarios so I've added there here so that the structure is the same for all of the scenarios.

Another aspect to notice is the modeling of time. Some modelers put Time as an actor, which I believe reflects faulty reasoning. Time is not an object from which one receives messages. Instead, it is a part of the infrastructure of the universe. Inside the system, we will ultimately have timer objects that provide explicit messages to indicate time outs, so that these timeout messages are shown as a "message to self." I find that a much more useful representation of time. Rhapsody provides a special icon for a timeout, a message to self that originates from a small box placed on the lifeline. To show a cancelled timeout, the message line is drawn with a dashed line. The hexagons in the figure are a UML 2.0 feature called condition marks, which shows the condition of the system. They are optional but they are included in the first two scenarios because I believe they clarify the scenario. Compare these scenarios with the third and make your own decision as to whether they clarify or obfuscate the scenario.

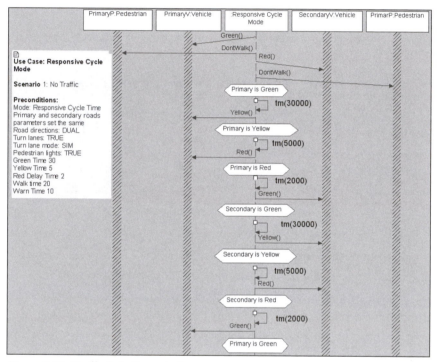

Figure 8.10 Traffic light Scenario 1

The first scenario is useful because it shows a very reasonable example of system behavior in the Responsive Cycle Mode: what happens when there is no traffic. A more interesting example is shown in the next scenario, shown in Figure 8.11.

The final scenario in this section occurs when a pedestrian comes from the secondary road when the primary road has green. This scenario is represented a bit differently than the previous two just to show the possibilities. First, the main lifeline in the middle is the «system» object rather than the use case. As previously mentioned, using the use case or the system object is a personal preference. Secondly, the conditional marks are not shown in the use case. Conditional marks can add clarity but they can also obscure the flow of the scenario if overused.

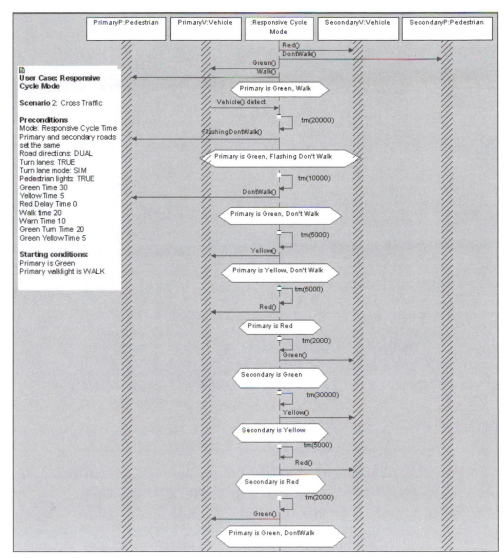

Figure 8.11 Traffic light Scenario 2

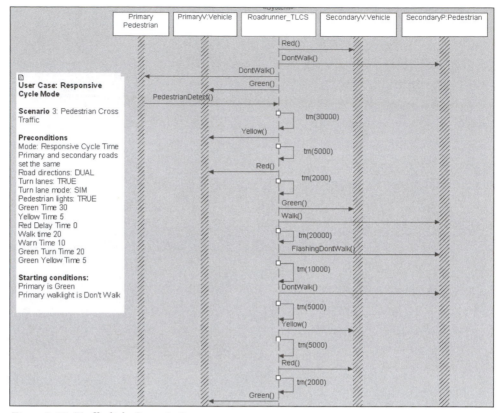

Figure 8.12 Traffic light Scenario 3

Answer 3.8 Operational View: CUAVS Optical Surveillance Scenarios

These scenarios for the Perform Optical Surveillance use case will seem a bit "high level" because we haven't identified the internal architecture of the CUAV and its payload. Nevertheless, we can still see how we expect the payload operator (in charge of managing the video camera on the CUAV) to interact with the system. In the next chapter, we will create a system architecture for the CUAVS and elaborate these scenarios to include that level of detail. For now, however, the CUAVS is a black box that interacts with various identified actors.

The first scenario, in Figure 8.13, shows manual movement of the camera in direct response to commands from the Payload Operator. First, the camera is enabled. Once enabled, the camera begins to transmit video at 30 frames-per-second at a resolution of 640 × 480. The command processing to move the camera is independently

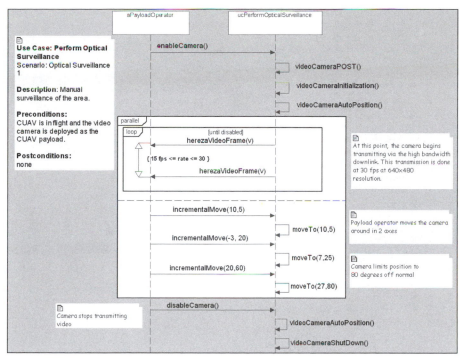

Figure 8.13 CUAV Scenario 1

processed from the image acquisition. This is shown with the parallel interaction fragment operator. The image acquisition occurs repeatedly until the camera is disabled, as shown with the loop interaction fragment operator.

In the second scenario, shown in Figure 8.14, automatic tracking of a target is shown. In this example, the Payload Operator moves the camera around until an "interesting" area is shown. He then selects a portion of the visual display and commands the system to identify targets within that area. The system identifies two targets (something that looks like a tank and something that looks like a building). These are highlighted in the visual frame and the potential targets are given target IDs. The operator then selects the target ID for the tank and commands the system to track it. As the tank moves around, the camera automatically adjusts to follow the tank's movements. At the end of the scenario, the operator commands the system to return to manual tracking operation.

The last scenario, shown in Figure 8.15, involves the operator selecting an area, centering it, and then zooming in. Once zoomed in 300%, he pans around, eventually deciding there is nothing of interest and so resets the zoom level to normal (100%).

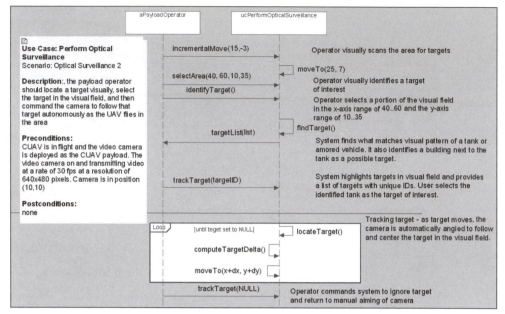

Figure 8.14 CUAV Scenario 2

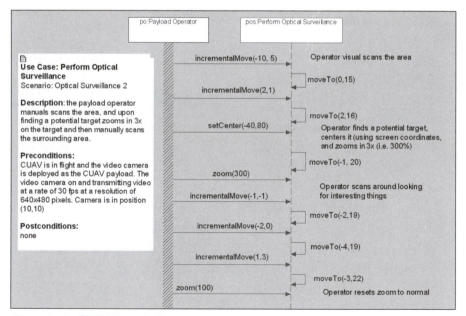

Figure 8.15 CUAV Scenario 3

Answer 3.9   Specification View: Use-Case Descriptions

This problem should pose no difficulties for the reader. The purpose of the problem is just to provide an opportunity to actually create a use-case description. Use-case descriptions are important because they provide necessary documentation for the understanding of the use case. Remember, just because you're building a model doesn't mean that you don't need description and comments!

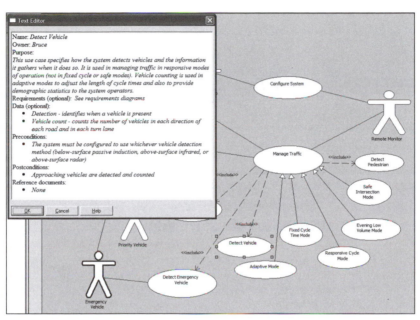

Figure 8.16 Roadrunner use-case description

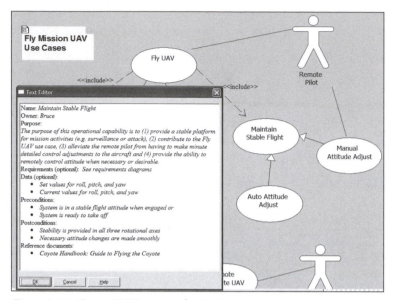

Figure 8.17 Coyote UAV use-case description

Answer 3.10 Specification View: Capturing Complex Requirements

The first part of this problem should be straightforward and is meant to be a gentle introduction to the power that state machines bring to the requirements-capturing process. This behavior can be expressed with three states—one to provide a delay before you start the flashing (to make sure vehicles in the intersection when this mode is initiated can clear the intersection), one for flashing "on" and one for flashing "off". The transitions between the states are all based on time.

The second state machine is significantly more complex. In order to better understand it, it will be incrementally constructed using the hint provided in the problem statement. The incremental approach—solve one part of the problem and get it to work before adding more complexity—is one that I recommend for all modeling, but especially when the going gets complex. Too many modelers fail in their efforts because they wait too long before they try to get the model executing. By solving the problem in small steps and demonstrating that the solution works via execution, we are implementing the "nanocycle" of the Harmony process, described in Chapter 2. This enables us to achieve a correct solution much more efficiently than the more common OGIHTW approach to modeling.[2]

---

[1]   Oh God I Hope This Works. In my experience, prayer might be a wonderful thing, but it is a suboptimal engineering approach!

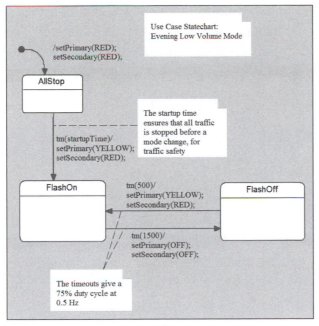

Figure 8.18 Use-case state machine 1

The first step is to get the basic fixed cycle mode behavior working. This is shown in Figure 8.19. See, now, that's not too hard, is it? While working this problem, I first did the statechart in the figure without and-states, using the same state names

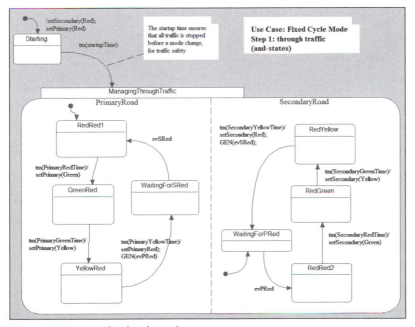

Figure 8.19 Basic fixed cycle mode

as in the figure. Although the no-and-state version was slightly simpler, it appeared that it would be more difficult to incrementally add the turn lane and pedestrian traffic later, so I decided to go with the and-state version. Notice that the two and-states coordinate on the basis of propagated events. A propagated event is an event generated in one statechart (or and-state) as a result of accepting an event. The latter event is then consumed by the other statechart (or and-state).

The next step is to add turn lanes. While I'll show this as a fait accompli, in reality I constructed this by first adding the latches, executing that version, and then adding the turn lanes.

To manage the complexity of the fixed cycle mode with the turn lanes added, I used submachines, one for the primary traffic and another for the secondary. The top level state machine for this is shown in Figure 8.20. The next two figures, Figure 8.21 and Figure 8.22, show the submachines for the composite states shown in the figure. These submachines both contain and-states to concurrently manage the through and turn traffic and the latch for the turn lane. The and-states ensure that the nested state machines execute independently except for explicitly stated synchronization points. These occur with a combination of propagated events (as in the first stage of the solution) and the use of guards to take the appropriate transition branch if there is turn traffic waiting.

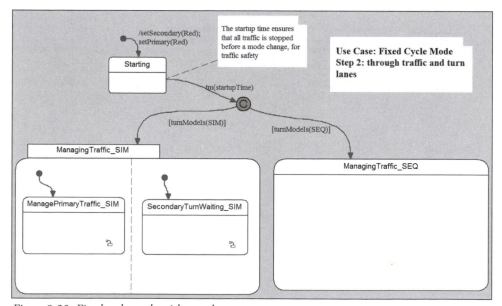

Figure 8.20 Fixed cycle mode with turn lanes

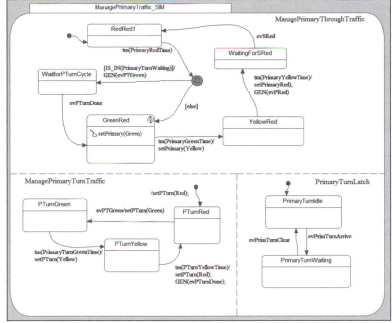

Figure 8.21 Fixed cycle mode primary submachine

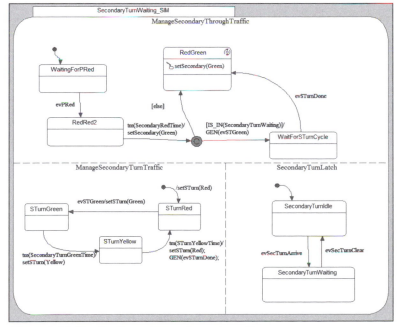

Figure 8.22 Fixed cycle mode secondary submachine

Once this model executes properly, I can move on to add the remaining features. In the last step, I added the pedestrian traffic management. This is shown as a set of diagrams. Submachines are used to manage the complexity by distributing it across multiple diagrams. The high-level view is the same as Figure 8.20. The submachine for the Manage Primary Traffic is shown in Figure 8.23. The processing for the primary turn and pedestrian traffic is specified in the submachines defined in this state.

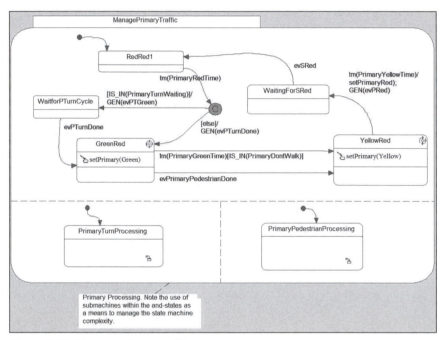

Figure 8.23 Manage primary traffic

The turn processing is the same in step three; this is shown in the submachine for the Primary Turn Processing State, in Figure 8.24.

And finally, the primary pedestrian processing is shown in Figure 8.25. It uses the same latch pattern as the turn light processing and communicates with the primary through traffic processing by sending the evPrimaryPedestrianDone event, which enables the green through light to go to yellow.

The secondary traffic works much the same as the primary. Note that while the state machines look (and, to be fair, are to some degree) complex, it is nevertheless much clearer than the equivalent voluminous text of equal completeness and accuracy would be.

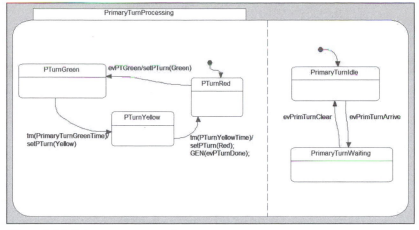

Figure 8.24 Primary turn processing

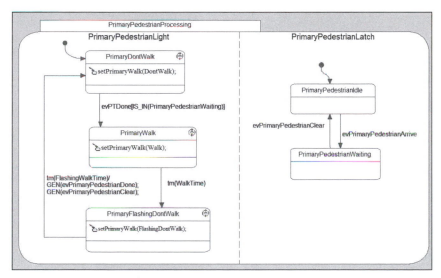

Figure 8.25 Primary pedestrian processing

Additional Questions

- Fill out the statecharts for the SEQ (sequential) turn mode using the statecharts for the SIM mode.

- Try doing the SEQ model solution without the use of and-states.

Answer 3.11 Operational to Specification View: Capturing Operational Contracts

In this exercise, you were given several distinct tasks to accomplish:

1. Draw the block diagram representing the use case, the actor(s), and the interfaces between them.

2. The messages that come from the actor(s) must be collected into provided interfaces on the appropriate port of the use-case block and required interfaces on the actor. Messages sent to the actor(s) must be collected into required interfaces on the use-case block and provided interfaces on the actor port.

3. For each service (operation) the system provides, specify pre- and post-conditions and the types and ranges of parameters, if necessary.

4. Construct a use-case activity diagram representing all of the scenarios previously specified for the use case.

5. Define a state machine for the use-case block that is consistent with the set of scenarios.

Step 1: Draw the Block Diagram

Figure 8.26 shows the block diagram representing the Perform Optical Surveillance use case. Note that the use case and actor have both become blocks (objects), complete with ports and interfaces. The use-case block provides one interface,

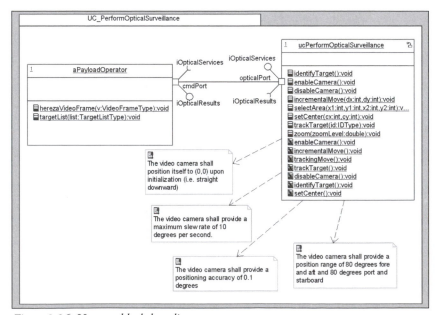

Figure 8.26 Use-case black-box diagram

iOpticalServices; this same interface is required by the actor. Similarly, for messages coming from the use-case block to the actor, another interface, iOpticalResults is required by the use-case block and provided by the actor. Interfaces used in this way are called interface conjugate pairs; that is, when the interfaces offered by one port are required by the other, and vice versa.

Remember that the interfaces specify the services provided or required by some element. By providing a set of services in the iOpticalServices interface, the use-case block is declaring that it promises to perform those activities. By requiring the other interface, iOpticalResults, it declares that it needs the actor to provide those services.

The operations are shown inside the blocks with a key symbol used to indicate "protected" visibility and no symbol to indicate that they are public.

Step 2: Define the Interfaces

The interfaces themselves are shown in Figure 8.27. Of course, they only contain the public operations for the use-case block and the actor. Specification of the interfaces was a straightforward job of selecting the public operations defined on the block and entering them as part of the interface.

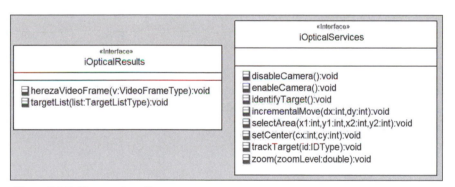

Figure 8.27 Use-case interfaces

Step 3: Defining Pre- and Post-Conditions for the Interfaces

The next step is to iterate over each of the operations in the interface and specify the pre- and post-conditions. Pre-conditions are things that must be true prior to the invocation of the service. Post-conditions are things that are guaranteed to be true by the service provider upon completion of the service.

The common repository for such information is in the description field provided by Rhapsody for each operation. However, it is more parsimonious for our purpose here to collect them all in a table (See Table 8.1).

Table 8.1 Interface pre- and post-conditions

Interface	Service	Pre-condition	Post-condition
iOpticalResults	herzaVideoFrame	Video camera is operating	Frame is stored in frame buffer for viewing and analysis
	targetList	Targeting enabled; identifyTarget command previously sent	List of targets is stored for analysis
iOpticalServices	disableCamera	None	Camera is returned to starting position and shut down
	enableCamera	None	Camera is powered and a Power On Self Test is performed, If errors occur, then error codes are sent; otherwise, the camera is initialized and centered.
	identifyTarget	Camera is operating and search area or route has been specified.	System identifies a list of targets, each with a unique target ID.
	incrementalMove	Camera is operating.	If the incremental move is within the range of the gimbaled camera mounting, then the camera is moved incrementally +x units and +y units. If the commanded position is out of range, then the camera will move to the closest valid position.
	selectArea	Camera is enabled	If the area is within the limits of the system, then the search area is defined. If not, then an error is sent and the closest matching search area will be selected.
	setCenter	Camera is enabled	If specified point is in range, the point is set to the be the search center. If not, an error is returned and the center remains unchanged.
	trackTarget	Camera is on and potential targets have been tagged with IDs	If the target ID is known to the system, then it will track that target within the center of the visual field with a combination of flight maneuvers and camera gimbal movement, depending on the state of the CAUV.
	zoom	Camera is on.	If the zoom parameter is within bounds, then the camera shall zoom in or out, auto adjusting focus. If the zoom level is out of range, then the camera shall clip at the minimum or maximum zoom, whichever is closer. The units are percentage, so that a parameter value of 300 means 3x magnification, 50 means 0.5 magnification.

Step 4: Construct a Use-Case Activity Diagram Representing All of the Scenarios Previously Specified for the Use Case

It is useful to have an overview of all the different scenarios at once. Since a sequence diagram represents a particular scenario, how can this overview be represented? Although in UML it is theoretically possible with the interaction fragment operators to represent all branch and concurrent points, it is not generally practical to do so. One easy way is to use an activity diagram to sum up all the messages that can occur. If two scenarios depart at some particular branch point, this can be easily represented using the activity diagram branch operator.

For the set of scenarios defined for this use case, the derived activity diagram is shown in Figure 8.28.

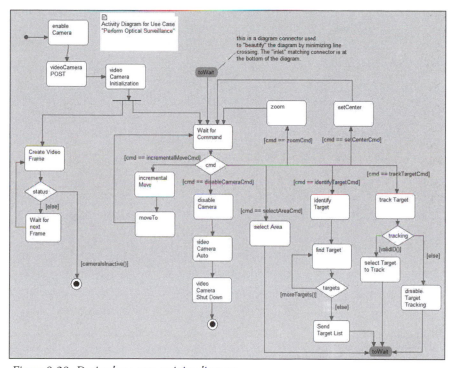

Figure 8.28 Derived use-case activity diagram

Step 5: Define a State Machine for the Use-Case Block that is Consistent with the Set of Scenarios

The last task from the problem statement was to construct a statechart for the use-case block that invoked the behaviors under the right circumstances, at the right time, and in the right order. Note that the activity diagram is recommended to be a documentation-only thing and is used to specify the user case per se; however, the statechart is executable and specifies the use-case block. The advantage of using a statechart over an activity diagram is that the former can use events such as timeouts, arrival of commands, etc. to trigger transitions. This isn't possible in UML 2.0 activity diagrams.

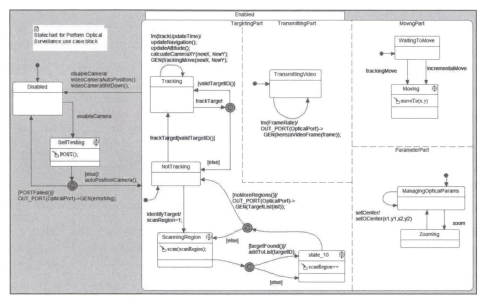

Figure 8.29 Use-case block statechart

A powerful aspect of using statecharts in this way (namely, to represent scenarios) is that the statechart fully specifies the allowable sequences and shows which sequences are disallowed. A single scenario cannot do that, but a statechart can. In this case, shown in Figure 8.29, we show logical concurrency with and-states.[3] Thus, we see that tracking is independent from, for example, moving of the gimbaled camera and managing optical parameters such as the zoom level. How target tracking takes place is also much more explicitly stated in the statechart than in the more informal sequence diagrams.

References

[1] Douglass, Bruce Powel, *Real-Time UML, Third Edition: Advances in the UML for Real-Time Systems*, Addison-Wesley, 2004.

[2] Douglass, Bruce Powel, *Doing Hard time: Developing Real-Time Systems with UML, Objects, Frameworks, and Patterns*, Addison-Wesley, 1999.

---

3   However, this does not imply or constrain the internal concurrency architecture. Here we are capturing things that are independent in their order of execution but there are an infinite number of ways that can map to a set of internal OS threads.

Systems Architecture: Answers

Answer 4.1 Organizing the Systems Model

This exercise has two parts, one for each of the two systems we're working with in this book. The first answer shows the organization of the Roadrunner Traffic Light Control System. This is shown in both model browser and package diagram views.

First, in Figure 9.1, we see a package diagram showing the set of packages used to organize the system as a whole. The mission for each package is stated in a brief comment. In the next figure, Figure 9.2, the details of the systems model area are shown. Note that the package for the system elements is named "_System". This is so that it comes first in alphabetical order in the browser view.[1]

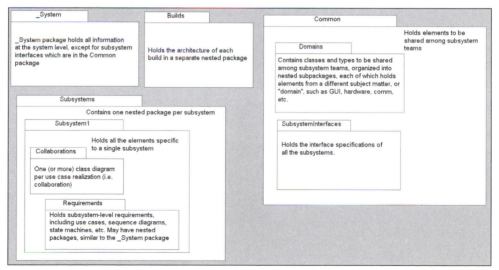

Figure 9.1 Roadrunner overall model organization

---

[1] This isn't strictly necessary in Rhapsody because it gives you control over package ordering in the browser, but it is a personal habit of mine.

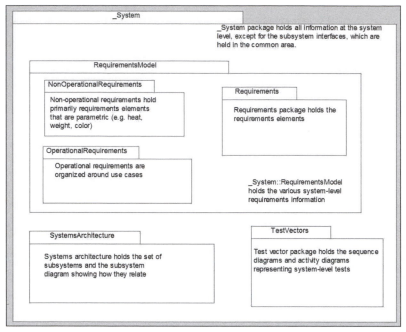

Figure 9.2 Roadrunner systems area organization

The Requirements package nested inside the RequirementsModel package of Figure 9.2 holds the individual requirements statements that trace to use case or more detailed operational elements; these requirements statements are represented as SysML requirements elements. The OperationalRequirements package holds the use cases and their detailed specifications with state machines, activity diagrams, and sequence diagrams. The NonOperationalRequirements package holds requirements that trace to the System object and other nonoperational constructs.

The next two figures show the same information but in the browser tree format. The latter is preferable for dynamically navigating through your model but both views are useful.

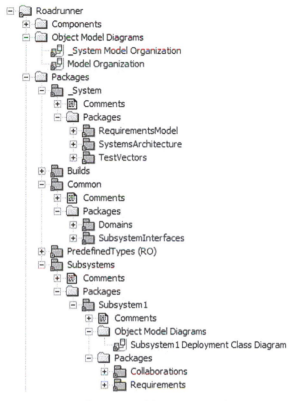

Figure 9.3 Roadrunner model organization browser view

The second part of the exercise is to structure the Coyote UAV model into a set of interacting models along the guidelines set out in Chapter 4. The Coyote UAV system is two or three orders of magnitude larger in scope than the Roadrunner system and, in order to be effective, it must be managed in multiple models:

- Systems Model

- Common Model

- (n) Subsystem Models

The systems model will be used for system level requirements and architecture capture. This model has separate packages for subsystems because it is those very packages that will be handed off to the subsystem teams when the project transitions from systems engineering to subsystem development. These packages hold the "software and hardware specification" models that will serve as the starting basis for the elaborated subsystem models. They will hold only requirements and requirements-related elements. The internal structure of the subsystem models will be handled by the subsystem teams in the subsystem models directly.

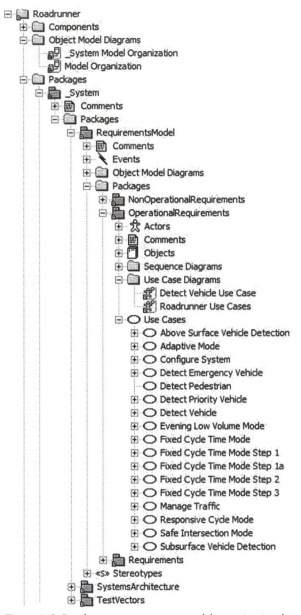

Figure 9.4 Roadrunner systems area model organization browser view

The common model will hold the elements to be shared among the subsystem teams, such as the subsystem interfaces and class specifications that span multiple domains (more on domains later in this chapter). Lastly, for each subsystem team, there is a separate subsystem model that can be thought of as a systems-model-in-the-small. That is, the subsystem model is organized along the same lines and using the same principles as the system model. It has a package for subsystem requirements and a package for the internals of the subsystem, specified primarily as collaborations among the objects held within the subsystem.

Central to the notion of decomposition of a system into a set of cooperating models are the notions of how to share these more-or-less-independent-but-collaborating models. Models may be shared either by value or by reference. When you share a model by value, the client imports a copy of the model. The client is then free to manipulate the copy, and add and delete elements to that copy without it affecting the original. If we share a model by reference, then the client copy is read-only and changes in the original model are reflected in the imported copy automatically. In Rhapsody, both are done with the "Add to Model" feature. When invoked, you are permitted to load an entire model, a set of packages, or an individual element, in either "by value" or "by reference" flavors. For models that you are expected to elaborate—such as the subsystem models—importing the model by value makes the most sense. In models where you want to reuse existing model elements, such as from a model that has classes to be used in multiple subsystems, importing the model by reference is most appropriate.

To this end, we've set up a set of models:

- Systems Model
- Builds Model
- Common Model
- (n) Subsystem Models

The Systems model is almost exactly the _Systems package in the Roadrunner solution. It differs primarily in that, while it has a package for each subsystem, the only contents of that package are requirements and requirements-related elements (use cases, etc.). It is not meant to contain the subsystem design—that is the job for the subsystem team to create in their own separate model.

The Builds model may be thought of as a "master" model in that it has very little of its own but imports all the subsystem models together to make a coherent executable instance of the system (aka, "prototype"). Early builds may be incomplete in the sense that not all subsystems are represented, the final electronic or mechanical hardware isn't used, or not all of the functionality of the various subsystems may be present. Nevertheless, even these early builds execute and can be validated. Over time, the later builds become increasingly complete and use more of the final hardware.

The Common model meets the same need as the Common package in the monolithic model solution, as a place where reusable elements can be placed so that they are accessible to the other models that wish to use them.

Lastly, the subsystem models are begun by importing the subsystem specifications from the Systems model. These models are then independently elaborated and detailed as their analysis and design processes proceed.

The CUAV Systems Model

The Systems model is shown in Figure 9.5 (package view) and Figure 9.6 (browser view).

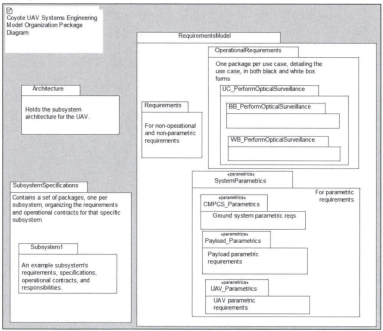

Figure 9.5 CUAV Systems model package diagram view

The Common model is shown in Figure 9.7.

Lastly, a sample subsystem model (only one, as we haven't yet identified the subsystem architecture) is shown in Figure 9.8. Of course, as we elaborate the architecture, there will be one of these subsystem models for each and every identified subsystem.

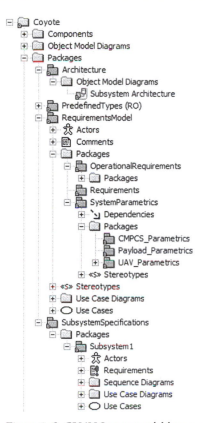

Figure 9.6 CUAV System model brower view

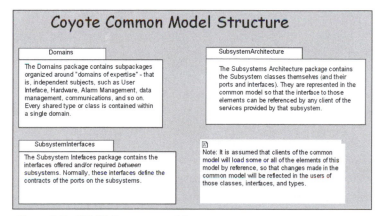

Figure 9.7 CUAV Common model

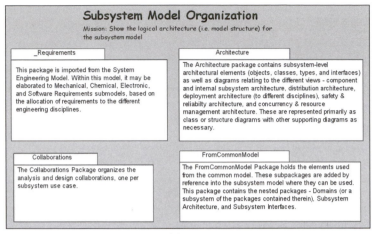

Figure 9.8 CUAV subsystem model

Answer 4.2   Subsystem Identification

The Roadrunner Traffic Light Control System is a fairly small system so the system architecture diagram is straightforward. I used a UML structure diagram to represent the system architecture. Note the use of stereotypes «System» and «Subsystem» to mark the architectural elements. This is completely optional but I find it useful to clarify the usage of elements in the model. Note that none of the elements shown in Figure 9.9 are "primitive"—they are all composites containing a possibly rich set of internal parts.

The ports between the Roadrunner subsystems are not (yet) typed by interfaces, but the "need lines" among the elements are straightforward. One can certainly imagine these connections are required for system operation and with a little thought one could definitely define a set of services provided and required across these ports. The definition of these interfaces is dealt with in an exercise later in this chapter.

Notice the set of subsystems for the traffic light control system in Figure 9.9. There is a different subsystem for the primary vehicle light assembly (P_Vehicle-LightAssembly) than for the secondary road (S_VehicleLightAssembly). The two subsystems are most likely identical in their structure and behavior but play different roles. The model shown uses a multiplicity of 1 with each, but an alternative representation would have been to have a single subsystem part for this role called VehicleLightAssembly and have a multiplicity of 2. I selected the former approach because the roles between the two subsystems are different to their client (the Inter-

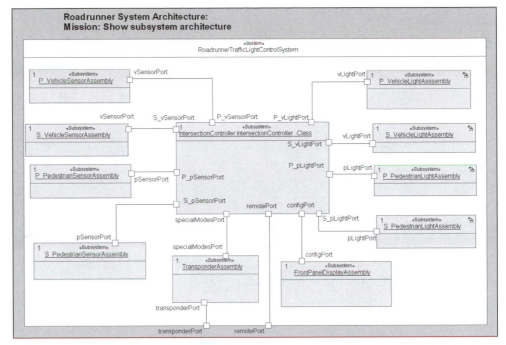

Figure 9.9 Roadrunner system architecture diagram

section Controller) and using different connections made that distinction clearer, but either approach could be used.[2]

The Coyote Unmanned Air Vehicle is about two orders of magnitude larger in scope than the traffic light control system. For this reason, the architecture is depicted on a set of diagrams rather than a single diagram. The first of these, Figure 9.10, shows the interconnected systems comprising the CUAV conglomerate. We can see that there are three kinds of systems identified: the ground systems (for mission planning and mission control), the CUAV vehicle, and the payloads. There are four different payloads shown: missiles, forward-looking infrared (FLIR), video, and synthetic aperture radar (SAR). Each of these is stereotyped «System» and will be decomposed into separate subsystem diagrams for each. While it is possible to put all this information in a single diagram, it makes that diagram far more difficult to comprehend.

Figure 9.10 is a class diagram, rather than the slightly more restrictive structure diagram, just to show an alternative means to depict architecture. Since it depicts

---

2   As a general guideline, when a set of objects plays the same role with respect to a subject object (they are treated more or less as the same), then I use a single association with * multiplicity. When a set of objects fulfill different roles, then I use a different association (to the same class) for each role.

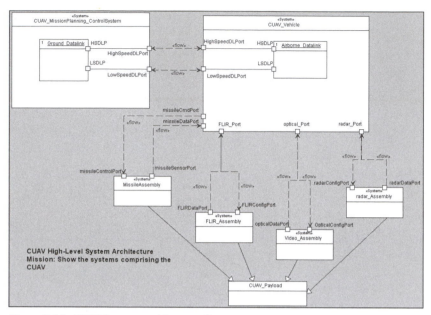

Figure 9.10 CUAV system architecture diagram

classes, not objects, links cannot be drawn between the ports.[3] Therefore, information flows are shown connecting the ports. The same contents shown in a structure diagram are shown in Figure 9.11, but in this case the primary architectural elements are parts of the overall project, and so can be linked with connectors.

The enclosing structure CUAV_Project is stereotyped «SystemOfSystems» to indicate that it is "larger" than the contained systems. Because this is a structure diagram, the internal parts are instance roles, and so can be linked via connectors, rather than having to reply on information flows. We can also indicate multiplicity on the parts. The missile_Assembly, FLIR_Assembly, video_Assembly, and radar_Assembly parts are limited to at most one per vehicle but have * multiplicity because there may be multiple vehicles active in the same system configuration.

In the previous figures, the payloads were assigned «System» stereotypes and are modeled at the same level of abstraction as the vehicle and the ground station. A case can be made for making them subsystems of the vehicle. After all, they are mounted on and deployed from the vehicle. While it would certainly not be incorrect to model the payloads in that way, in this case I did not. The reason is that the

---

3 Links can only be drawn between objects and between ports on objects; associations can be drawn between classes, but not between ports in UML 2.0. A link is, of course, an instance of an association. Links could have been drawn between the elements on the diagram had the classes on the diagram been instantiated into objects.

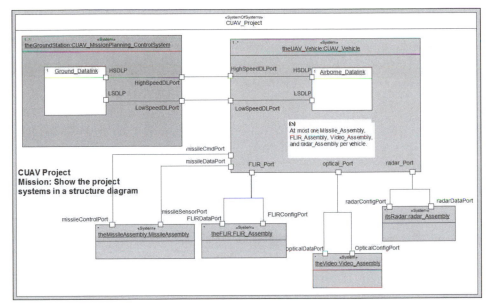

Figure 9.11 CUAV system architecture as structure diagram

payloads are removable, replaceable, and possibly mountable on different vehicles, some of which may not even be within this project. To make their reuse even easier to conceptualize, I decided to make them separate systems, rather than subsystems of the vehicle.

The systems defined within the CUAV project must be decomposed into their primary subsystems as well. This is done in the following set of diagrams. Figure 9.12 shows the subsystem architecture of the ground station. Note that the Ground Datalink subsystem and the Data Storage subsystem are singletons (i.e., have a single instance) within the ground station but there are up to four manned stations, and each of these has two stations within them, one for controlling a UAV and one for receiving and processing the reconnaissance data. The association among the manned stations allows them to be slaved together into a single station for low-vigilance monitoring.

The guidelines used to link together the subsystems are simple. If one subsystem requires services or data from another, a link is emplaced between them. All links (at the subsystem level) employ ports as a means to help ensure strong encapsulation.

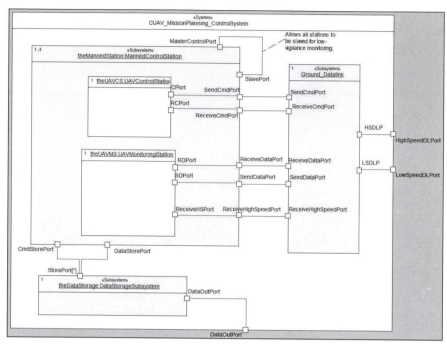

Figure 9.12 CUAV ground station subsystem architecture

Figure 9.13 shows the vehicle itself. It is organized into a set of subsystems as well. The subsystems are summarized in Table 9.1.

Table 9.1 CUAV subsystem summary

Vehicle Subsystem	Links with	Purpose
Airborne Datalink	High-speed external port Low-speed external port Flight Management Fire Control Attitude Control Engine Control Navigation Reconnaissance	Links vehicle to ground station for both low-speed (command) data and status and high-speed data. Also unmarshalls messages received over the communications links and delivers them to the appropriate subsystem.
Attitude Control	Airborne Datalink Mechanical Hydraulics Flight Management	Controls and monitors the attitude (roll, pitch, yaw) of the vehicle by commands position changes in the vehicle's control surfaces.

Table 9.1 CUAV subsystem summary (continued)

Vehicle Subsystem	Links with	Purpose
Engine Control	Airborne Datalink Flight Management	Controls engine output and monitors engine status.
Fire Control	Airborne Datalink Targeting Missile Command external port Missile Data external port	Manages on-board weapons systems, both status and control, including the delivery of fly-by-wire and fire-and-forget Hellfire missiles.
Flight Management	Airborne Datalink Attitude Control Engine Control Mechanical Hydraulics Navigation Fuel Management	Manages general flight, which can be fully or partially automated or may be managed completely from the ground.
Fuel Management	Flight Management	Monitors fuel status and shunts fuel as necessary for weight balance.
Mechanical Hydraulics	Attitude Control Flight Management	Provides direct control over air control surfaces—wings, rudders, elevators, and ailerons. Also used to control landing gear.
Navigation	Airborne Datalink Flight Management	Used for control and monitoring of position in the airspace. Also contains navigation data and tools, such as digital maps and flight plans.
Reconnaissance Management	Airborne Datalink Targeting FLIR external port Optical external port Radar external port	Gathers and processes reconnaissance data for both on-board target tracking and for remote surveillance (transmitted to the ground station).
Targeting	Fire Control Reconnaissance Management	Identifies and tracks targets of interest.

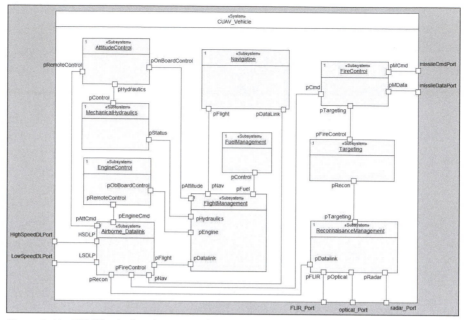

Figure 9.13 CUAV vehicle subsystem architecture

The payloads are not broken down into their subsystems here. One reason is that it is likely that a single team will be developing each of the payloads, so it is less crucial to do that decomposition early. Secondly, the payloads are clearly less complex than either the ground station or the vehicle.

One thing to note about the system-to-subsystem decomposition performed in this answer: the subsystems themselves are multidisciplinary. That is, each subsystem contains software, electronic, mechanical and possibly chemical engineering aspects. That decomposition is discussed in a later problem in this chapter.

Answer 4.3   Mapping Operational Contracts into the Subsystem Architecture

In the process of developing systems, the tradition is to first create a set of requirements. In the MDA/UML world, these requirements are organized into and around use cases. Next, we define a system architecture. This has been addressed for both sample problems in previous problems in this chapter. The next step is to map the requirements into the various subsystems. This consists of two primary steps:

1. We must map the operational contracts (services) into the internal architecture.

2. We must group these services at the subsystem level into use cases.

This problem deals with the first of these two steps. Conceptually it is a simple process.[4] A use case is realized (i.e., "implemented by") a set of collaborating system elements. At the architectural level these are systems, subsystems, and components. Each service at the outer-most system level ("system" level for the Roadrunner Traffic Light Control System, and "system of systems" level for the Coyote UAV) is decomposed into services provided by the various subsystems. In this step of the process, we decompose the system-level services into subsystem-level services. We represent these service invocations primarily as sequence diagrams messages.

Mechanically, the decomposition of the system services can be done "in place" or by decomposing the use-case lifeline. By "in place," we mean that the original sequence diagram(s) are copied and the subsystem lifelines and messages are added. This is only useful for small systems. For larger-scale subsystems, or systems with more parts at the next level of abstraction down, it is better to use the lifeline on the original sequence diagram as a reference to a more detailed sequence diagram; this latter sequence diagram represents the interaction of the lifeline's internal parts in the very same scenario. This is a new UML 2.0 feature and is one of the most important additions to the UML. In Rhapsody, creating the sequence diagram representing the decomposed lifeline is easy—simply double click the lifeline to get the "Features" dialog, and then select <new> in the "Decomposed" list control (or select the name of an already existing sequence diagram). This creates (or selects) the sequence diagram to be used as decomposition of the lifeline. To navigate to the referenced sequence diagram is easy—simply select the lifeline in question, right-click, and click on "Open reference sequence diagram."

When you decompose a sequence-diagram lifeline, the first question that arises is "How do I represent messages entering and leaving the sequence diagram?" The most common way is to use the system border, a special lifeline that represents all instances other than the ones explicitly shown. All messages entering or leaving the nested sequence diagram go to or from that system border lifeline. If you visualize the nested sequence diagram as being inside the boundaries of the lifeline on the original SD, this is the same as saying the incoming messages are coming from the "edge of the owner lifeline" and outgoing messages are sent to that same edge.

An alternative to using the system border is to copy the actors from the original sequence diagram. If all the elements sending or receiving events are actors, it is

---

4   Although, remember the Law Of Douglass that states "The difference between theory and practice is greater in practice than it is in theory!"

perfectly fine to copy them into the nested sequence diagram if desired. This is equivalent to the previous solution.

While other tools may work slightly differently, Rhapsody has an "analysis mode" for sequence diagram editing. This is useful for white-boarding scenarios since nothing is populated into the repository in this mode. When the scenario looks stable, the sequence diagram can be turned into a "design mode" diagram that commits the messages to the repository. In design mode, all the messages being received by a lifeline (representing a classifier) become operations (for synchronous messages) or events (for asynchronous messages) on that classifier.

First, let us work on the traffic light control system. The smarts in this system are centered largely in the Intersection Controller subsystem (see Figure 9.9). The first step is to take the sequence diagram for Scenario 2 and decompose the Responsive Cycle Mode use case. That creates an empty sequence diagram that is referenced from the original lifeline. Then open the new sequence diagram and drag the elements (parts of the Roadrunner Traffic Light Control System object) onto the diagram to create the lifelines. Add a system border lifeline for messages to enter and leave the sequence diagram. Now, working from the original, for every message coming from an actor to the use case lifeline, create a matching message from the system border lifeline on the nested diagram to the appropriate lifeline. Similarly, for every message from the use case lifeline to an actor, create a corresponding message from some element in the nested sequence diagram to the system border. Add internal messages among the elements as appropriate to glue the collaboration of these elements together. A nested sequence diagram for scenario 2 (Figure 4.3) is shown in Figure 9.14. Similarly, the more detailed view for scenario 3 (Figure 4.4) is shown in Figure 9.15.

So what does this mean to the subsystems? Once the messages are realized by the subsystems, corresponding services (operations and events) are added to the subsystems. In this particular case, we made the subsystems instances of classes, and those realized messages become operations defined on those classes. Once that is done, they can be added to the appropriate ports by collecting those services up into interfaces that specify the contracts of the ports. In Figure 9.16, the interfaces on the ports are not shown to minimize diagram clutter.

The traffic light controller is a relatively simple system, but the example illustrates the process of elaborating the services within the subsystems. In this case, the interfaces between the subsystems are not complex but this clearly allocates responsibilities to the architectural parts of the system. In Figure 9.16, I've added some additional operations that show up in some other scenarios (such as using

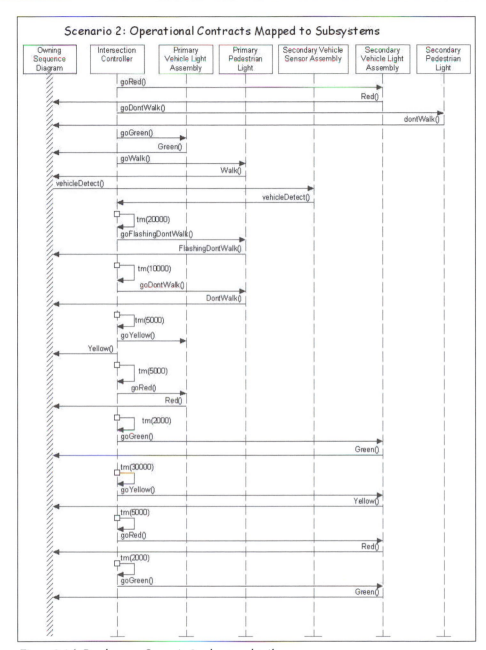

Figure 9.14 Roadrunner Scenario 2 subsystem details

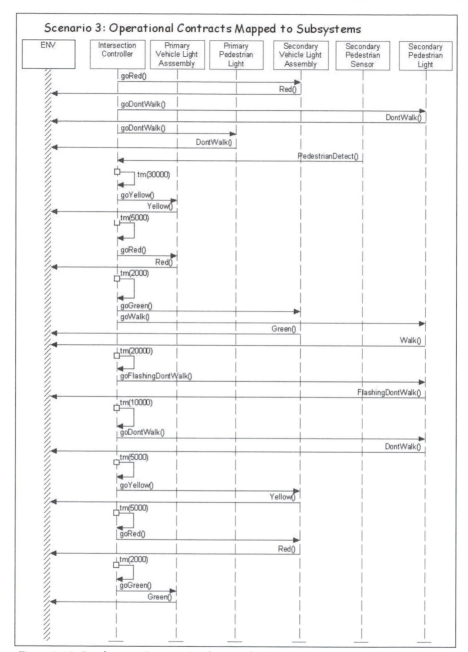

Figure 9.15 Roadrunner Scenario 3 subsystem details

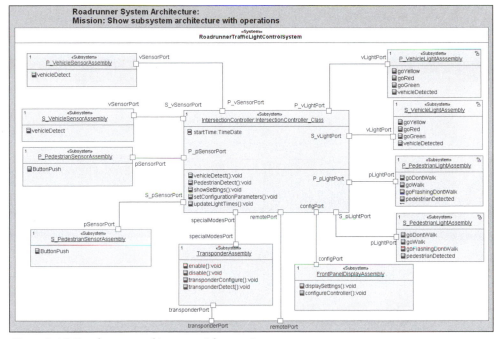

Figure 9.16 Roadrunner architecture with operations

the front-panel display for configuring the system and the transponder to detect emergency vehicles) not shown in this chapter, to round out the interfaces among the subsystems.

Let us now repeat the same process on the CUAV system. Of course, this is a much larger system so we will need to use both forms of sequence-diagram decomposition to manage the diagrammatic information.

We will start with looking at the interaction between the ground station (and its Ground Datalink), the aircraft (and its Airborne Datalink) and one of the payloads—in this case the video assembly. As stated in the original problem, we'll focus on the Perform Area Search use case and, more specifically, on Scenario 1 shown in Figure 4.7. Since that figure just references more elaborated sequence diagrams, we'll start with Figure 4.8, which details how the aircraft takes off. In the more detailed view shown in Figure 9.17, we will elaborate the Perform Area Search lifeline into a nested sequence diagram.

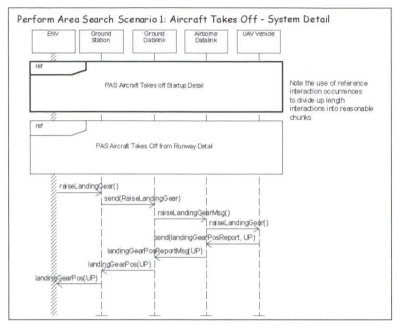

Figure 9.17 Perform area search Scenario 1 detail master

As with the previous example for the traffic control light, notice the messages coming in from the actors are summarized by the system border (named ENV in the figure) that connects the messages to the actors in the owning sequence diagram. Also notice, as mentioned before, the detail is too long to reasonably put into a single diagram, so interaction references are put into the scenario to hold much of the detail: the first shows the detailed interactions for setting up the aircraft, and the second shows the detail for taking off from the runway.

The first referenced interaction fragment is split across Figure 9.18 and Figure 9.19.

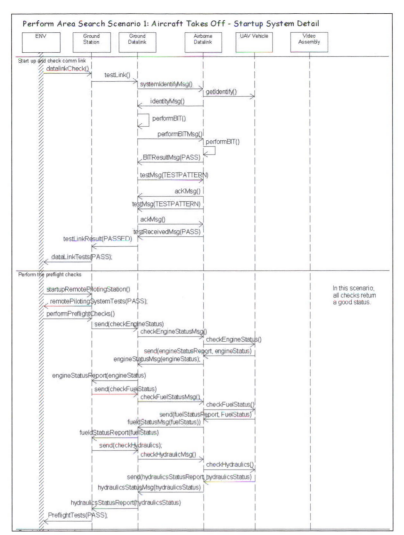

Figure 9.18 Setting up the aircraft detail (part A)

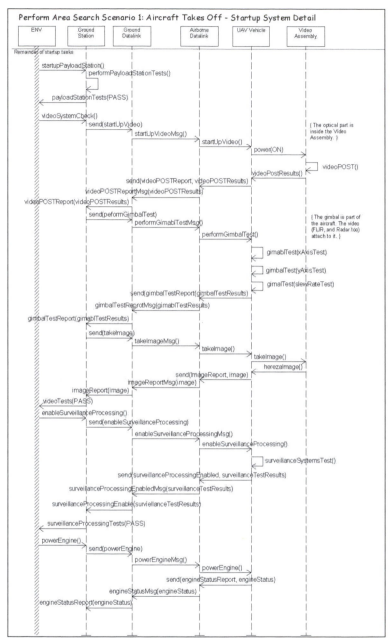

Figure 9.19 Setting up the aircraft detail (part B)

As implied in the previous diagram, showing the more decomposed level of detail tends to "explode" the information, so diagram and information management can become more difficult. This is where UML 2.0's decomposable sequence diagrams, which imply navigable links to the nesting levels, are very useful. If desired, you may add in hyperlinks to navigate backwards as well, but these must be added manually. Keep in mind that, because interaction fragments may be referenced in more than one place, you may need to create more than one such backward hyperlink. Such a hyperlink is shown (in a comment) in the next figure, Figure 9.21. If you click on that hyperlink in Rhapsody, the owning diagram opens. It is highly desirable to add such navigation links while constructing large-scale models. It not only greatly facilitates navigating through your model, but also adding them as you add diagrams is far easier than adding them later as an afterthought.

As in the two previous diagrams (which are a single diagram in the tool) this diagram is broken up across two figures as well for formatting purposes. In this case, the breakpoint for splitting the diagram is between the two parallel regions of the diagram. This is indicated with the dashed line within the parallel region of the figure.

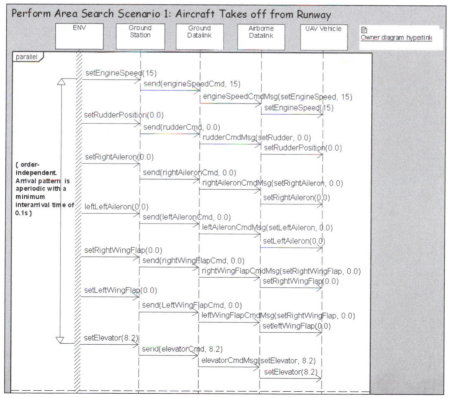

Figure 9.20 Aircraft takes off from runway detail (part A)

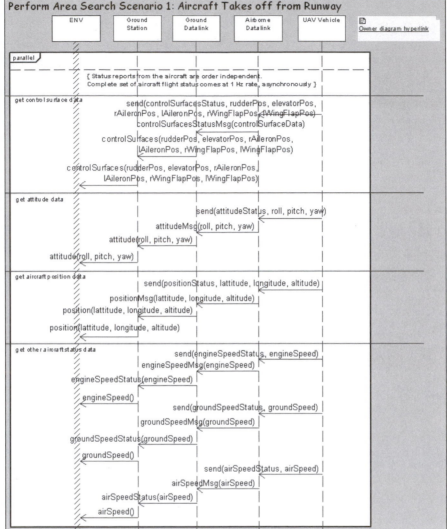

Figure 9.21 Aircraft takes off from runway detail (part B)

Once the operations are allocated, as in the previous set of sequence diagrams, the allocation of operations can be shown in a structure or class diagram, as in the next figure, Figure 9.22.

Figure 9.22 is easily created by taking Figure 9.11 and toggling the specification/structured view to "specification" for the parts to show all the operations.[5]

---

[5]   Although you might need to set the Display Properties (on the right click menu) for the parts to select which operations you want to show.

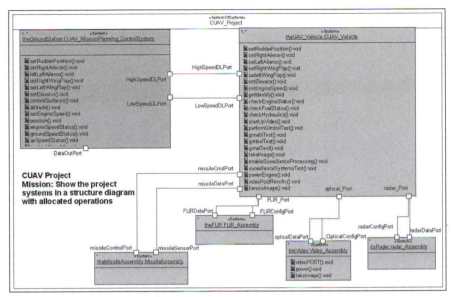

Figure 9.22 CUAV system allocated operations

Answer 4.4 Identifying Subsystem Use Cases

The first part of this exercise identifies the use cases for the Roadrunner Traffic Light system. However, we limit ourselves to just two system-level use cases—Detect Vehicle and Configure System. In a real project, we would, of course, decompose all of the system-level use cases and map them onto all of the subsystems. But, for this exercise, we limited ourselves to just two system-level use cases.

The first of these, "Detect Vehicle," has two specialized use cases: "Above Surface Vehicle Detection" and "Subsurface Vehicle Detection." Given the subsystem architecture of the traffic light controller, it is clear that most of the action for these use cases will take place in two subsystems, the Vehicle Sensor Assembly and the Intersection Controller. Some of the requirements apply to both the specialized subsystems. Specifically, regardless of how the vehicle is detected, vehicle count and related statistics must be kept and managed. Further, all the detectors must be able to communicate over a wired bus. Therefore, in Figure 9.23, the "Detect Vehicle" system use case is decomposed into two subsystem-level use cases, "Manage Vehicular Traffic Statistics" (mapped to the Intersection Controller subsystem) and "Communicate via Wire" (mapped to both the Intersection Controller and the Vehicle Sensor Assembly subsystems). The decomposition of the use cases is represented with the «include» dependency. The mapping is indicated in the figure with constraints; in

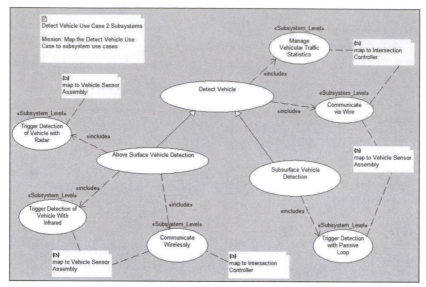

Figure 9.23 Roadrunner detect vehicular use case maps to subsystems

addition, the use cases reside in the relevant subsystem areas of the model. Use cases shared by multiple subsystems can be stored in the system area, in the Common package, or replicated in the targeted subsystems, as desired.

The specialized use cases are likewise decomposed into subsystem-level use cases with the «include» dependency.

The Configure System use case is similarly decomposed to a set of smaller use cases, each of which maps to a specific subsystem. The only exception is the Configure Wireless use case is system level, but is decomposed into three part use cases that are assigned to subsystems.

A couple of things to note about Figure 9.24. First, the "Configure Operational Modes" is assigned to the Intersection Controller subsystem. This implies that all the more specialized forms of this use case, such as "Configure Emergency Mode," are mapped to that subsystem as well. Secondly, the more specialized forms for configuring the wireless networks are mapped to the transponder assembly, the vehicle sensor assembly, and the intersection controller. Remember that the transponder receives signals from priority and emergency vehicles while the vehicle sensor assembly may have either wired or wireless connection to the intersection controller. Therefore, each of the systems must be able to configure its wireless communications.

The latter part of the exercise takes the CUAV system and maps the Manual Attitude Adjust and Perform Area Search use cases to the subsystems. We show the mapping of the Manual Attitude Adjust use case to the subsystems in Figure 9.25.

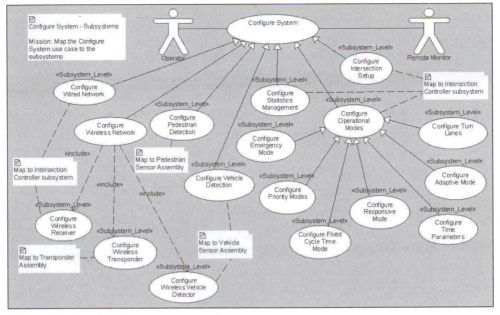

Figure 9.24 Configure use case maps to subsystems

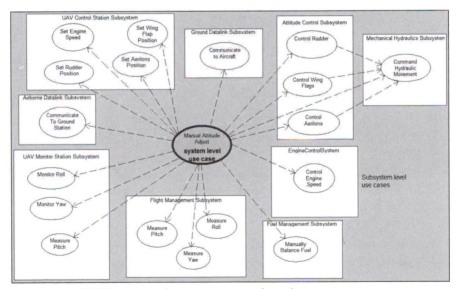

Figure 9.25 CUAV manual adjust use case mapped to subsystems

In this figure, we use a different means to depict the allocation of use cases to the subsystems. In the previous example, we used constraints to show the allocation. In Figure 9.25, we show the allocation with system boundaries, one per subsystem.

I create a use-case diagram for each system-level use case to show the decomposition to subsystem-level use cases and their allocation to the various subsystems. I call this a "use-case allocation diagram," but it is simply a use-case diagram whose mission is to show how the system-level use case maps on to the subsystem architecture.

Figure 9.26 shows the same information as the previous figure, but now for the Perform Area Search use case. If you review the system use-case diagrams, you will find that Perform Area Search is a specialized kind of Execute Reconnaissance Mission use case, which is in turn a specialized kind of Execute Mission use case. That latter use case includes Fly UAV use case. Therefore, rather than directly represent all the included use-case mappings as well, Figure 9.26 shows the inclusion of the Fly UAV use case (which in turn includes the Manual Attitude Adjust use case previously shown).

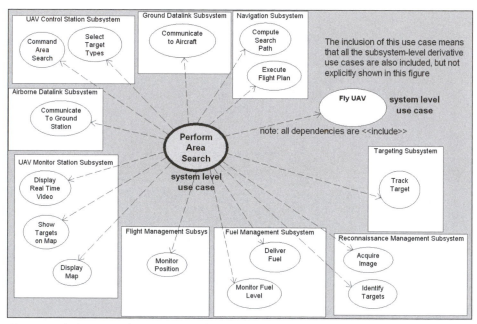

Figure 9.26 CUAV Perform Area Search use case mapped to subsystems

At the end of this allocation process, each subsystem has a set of use cases allocated to it. For example, the use-case diagram for the Navigation subsystem might look like Figure 9.27. In this figure, the actors are stereotyped «internal». This indicates that from the perspective of the subsystem, the elements are actors, but they are internal to the system being developed. That is, the peer subsystems become actors to the navigation subsystem. In the UML language, strictly speaking, actors are not merely a view of an object—they are its metaclassification. Therefore, Rhapsody requires the actors to actually be different elements than the subsystems they represent. For that reason, I preface the name with an "a" (for "Actor") to distinguish it from the subsystem of the (otherwise) same name, even though in usage, the "actor-ness" of an object is a result of perspective only.

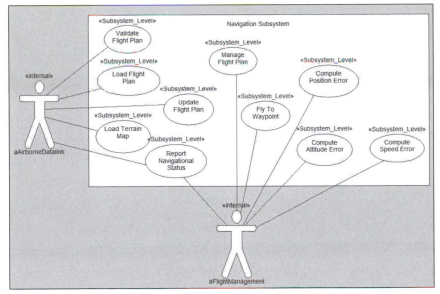

Figure 9.27 Navigation subsystem use cases

Also note that the use cases are stereotyped «Subsystem_Level» as well. This stereotyping is optional but allows the level of abstraction to be seen at a glance.

The next example, the flight management subsystem, has more use cases and certainly more actors. These are, again, derived from the system-level use cases, mapping a portion of the system-level use cases to the subsystem in question. Because some of the use cases (such as the Manage BIT[6] use case) connect to many actors, some of the actors are replicated. This means only that they appear multiple times on the diagram. This allows the diagram to minimize line crossing and simplify the diagram's appearance.

Lastly, once the decomposition to the subsystem use cases has been done, they can be dragged to the subsystem specifications area of the model. In the case of the Flight Management Use Case, the browser view of the Flight Management Subsystem package looks like Figure 9.29.

---

[6]   I use BIT here to stand for Built In Test, a common acronym for this purpose.

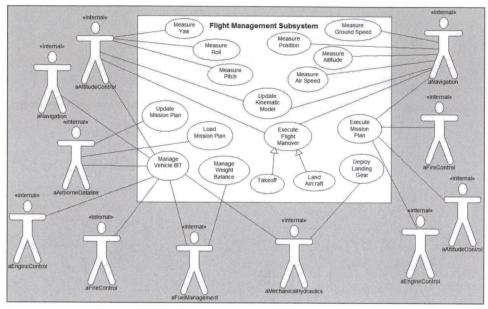

Figure 9.28 Flight Management Subsystem use cases

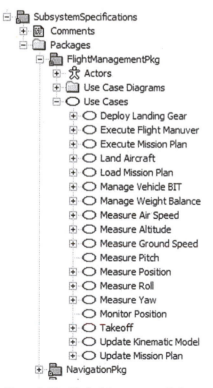

Figure 9.29: Flight Management Subsystem specification package contents

Object Analysis: Answers

Answer 5.1   Apply Nouns and Causal Agents Strategies

The strategy to apply is to underline the nouns and noun phrases. Below, the strategy has been applied.

The Vehicle Detector

Three <u>types of Vehicular Detectors</u> shall be supported: <u>subsurface passive loop inductors</u> (<u>SPLIs</u>), <u>above-surface infrared sensors</u> (<u>ASIs</u>) and <u>above-surface radars</u> (<u>ASRs</u>).

<u>Subsurface detectors</u> shall use a <u>wired interface</u> to communicate with the <u>controller</u>, while <u>ASIs</u> and <u>ASRs</u> shall support both <u>wired and secure wireless communication</u>. All <u>vehicle detectors</u> shall be able to perform <u>vehicle counting</u>.

In addition, ASIs and ASRs shall be able to receive <u>directional transmissions</u> from <u>priority vehicle and emergency vehicle transmitters</u>. The <u>maximum range</u> of such <u>reception</u> shall be no less than <u>250 feet</u> and no more than <u>1000 feet</u>.

<u>Figure 10.1</u> shows the <u>relevant measures</u> for both <u>ASI and ASR detectors</u>. When a <u>vehicle</u> enters the <u>detection area</u> (shown as the <u>shaded area</u> in the figure), the <u>detector</u> shall report the <u>presence of a vehicle</u>. <u>Separate detectors</u> are used for <u>each lane</u> in <u>each direction</u>.

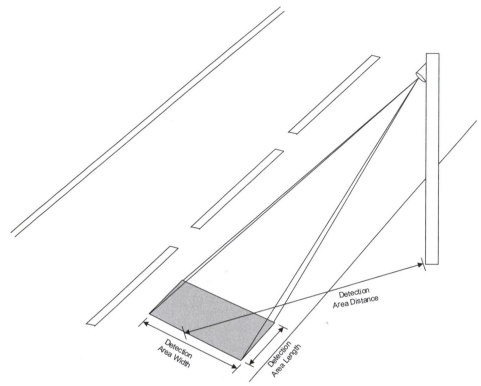

Figure 10.1 Infrared and radar vehicle detector

In Table 10.1, the noun phrases are categorized:

Table 10.1 Roadrunner Detect Vehicle nouns

Noun Phrase	Element Type	Element Name
types of vehicle detectors	class	VehicleDetector
subsurface passive loop inductors	class	PassiveLoopInductor
SPLI	class	PassiveLoopInductor
Above-surface infrared sensors	class	InfraredVehicleDetector
ASIs	class	InfraredVehicleDetector
Above-surface radars	class	RadarDetector
ASRs	class	RadarDetector
Subsurface detectors	class	PassiveLoopInductor
wired interface	class	WiredNetworkInterface
wired and secure wireless communication	unknown	unknown

Table 10.1 Roadrunner Detect Vehicle nouns (continued)

Noun Phrase	Element Type	Element Name
vehicle detectors	class	VehicleDetector
vehicle counting	attribute	VehicleCount.count
directional transmissions	class	Message
priority vehicle and emergency vehicle transmitters	actor	PriorityVehicleTransmitter EmergencyVehicleTransmitter
maximum range	UN[1]	useful for testing but not explicitly represented in collaboration
reception	class	Message
250 feet	UN	useful for testing but not explicitly represented in collaboration
1000 feet	UN	useful for testing but not explicitly represented in collaboration
Figure 5-1	UN	
Detection Area Width	attribute?	VehicleDetector.areaWidth
Detection Area Length	attribute?	VehicleDetector.areaLength
Detection Area Distance	attribute?	VehicleDetector.areaDistance
relevant measures	UN	
ASI and ASR detectors	class	AboveSurfaceVehicleDetector
vehicle	class	Vehicle
presence of a vehicle	event	evVehicleDetect
Separate detectors	class	VehicleDetector
each lane	attribute	VehicleDetector.lane
each direction	UN	

The second strategy is to identify the causal agents. In this case, the agents that ultimately cause events to enter into the system for the "Detect Vehicle" use case are the Vehicle actors (for a normal detection), and the Priority and Emergency Vehicle actors for the other kinds of vehicle detections. In our system structure, it will be sensors that identify these vehicles. For the normal vehicles, we have passive loop inductors that detect the metallic content of the vehicle and we have infrared and

---

[1] UN stands for Uninteresting Noun; in other words, something we're not going to explicitly model.

radar detectors that respond to heat and movement, respectively, within the range of the sensor. The specification also states that the infrared and radar detectors are connected via wireless links while the passive loop inductor is connected via a wired link. For the priority and emergency vehicles, the detection is done by receiving a directional wireless (IR, as it happens) transmission that indicates the approaching vehicle. The information about the detection will be used in different operational modes, but that is outside the scope of this strategy. However, this strategy has identified several kinds of vehicle detectors (passive loop inductor, infrared, radar, and wireless reception from a transponder or transmitter mounted on the vehicle). It has also identified wired and wireless interfaces to get that information to the internal system.

The results of applying these two object identification strategies are shown in Figure 10.2.

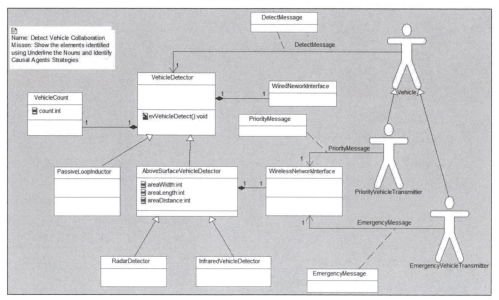

Figure 10.2 Vehicle Detect collaboration Strategy 1

For the second use case, "Fixed Cycle Mode," we'll employ the same strategies.

Mode 2: Fixed Cycle Time

Mode 2 is the most common operational mode. In this mode, the lanes cycle GREEN-YELLOW-RED in opposite sequences with fixed intervals. The system shall ensure that if any traffic light is non-RED, then all the lights for cross traffic shall be RED and pedestrian lights (if any) shall be set to DON'T WALK. Note that the <u>turn lane times</u> and/or <u>pedestri-an times</u> are only valid in this <u>mode</u> if (1) <u>the turn lane</u> and/or <u>pedestrian parameter</u> is set TRUE in the <u>RIC system parameters</u> and (2) if a <u>signal</u> from the <u>appropriate detector</u> determines the existence of <u>waiting traffic</u> for the <u>turn or pedestrian light</u>

The <u>durations</u> of the <u>light times</u> shall be independently adjustable by setting the <u>appropriate parameters</u> (see below). Note that in the <u>table</u>, the <u>values</u> in parentheses are <u>defaults</u>.

Table 10.2 <u>*Mode 2 parameters*</u>

Parameter	Value type	Description
Reset Parameters	FALSE, TRUE	(FALSE) Sets all the parameters for Mode 2 to defaults
Primary Green Time (PG2)	10 to 180 seconds	(30) Length of time the primary green light is on
Primary Yellow Time (PY2)	2 to 10 seconds	(5) Length of time the primary yellow light is on
Primary Red Delay Time (PR2)	0 to 5 seconds	(0) Length of time between when primary red light is turned on and the secondary green light is activated
Primary Walk Time (PW2)	0 to 60 seconds	(20) Length of time the primary WALK light is on when the primary GREEN light is activated
Primary Warn Time (PA2)	0 to 30 seconds	(10) Length of time the primary FLASHING DON'T WALK light is on after the WALK light has been on
Primary Turn Green Time (PT2)	0 to 90 seconds	(20) Length of time the primary turn light is GREEN. Note: only valid when the Primary Turn Light parameter is TRUE.
Primary Turn Yellow Time (PZ2)	0 to 10 seconds	(5) Length of time the primary turn light is YELLOW. Note: only valid when the Primary Turn Light parameter is TRUE.

The <u>default values</u> depend on the <u>system configuration</u>.

Table 10.3 Default cycle times for Mode 2

Turn Lane	Ped Signal	Green	Yellow	Red	Walk	Don't Walk	Turn Green	Turn Yellow
F	F	30	5	0	0	0	0	0
T	F	50	5	0	0	0	15	5
F	T	50	5	0	15	5	0	0
T	T	50	5	0	15	5	15	5

The values in Table 10.3 are true for each direction, independently. Thus, if the primary road has a car waiting in its turn lane and a pedestrian walking, but the secondary road has neither, then the following timing diagram represents the cycle times for simultaneous turn lane mode (i.e., the turn lanes in both directions for a road turn together and the straight traffic doesn't begin until the turn lanes have cycled to Red).

As before, let's categorize the noun phrases (see Table 10.4).

Table 10.4 Roadrunner fixed cycle mode nouns

Noun Phrase	Element Type	Element Name
Mode 2	State	System object
Operational mode	State	System object
Mode	State	System object
Lanes	Class	Vehicle Light Assembly
Sequences	Set of states	In various statecharts
Fixed intervals	Timeout transitions	In various statecharts
System	Object	System object
Traffic light	Class	Traffic Light Assembly
Lights	Class	Traffic Light Assembly
Cross traffic	Actor	Vehicle
Pedestrian lights	Class	Pedestrian Light Assembly
Turn lane times	Timeout transition	In statechart of Traffic Light Assembly
Pedestrian times	Timeout transition	In statechart of Pedestrian Light Assembly
Turn lane	Object	Instance of Vehicle Detector
Pedestrian parameter	Attribute	System configuration object
RIC system parameters	Class	System configuration object
Signal	Event	evVehicleDetect

Table 10.4 Roadrunner fixed cycle mode nouns (continued)

Noun Phrase	Element Type	Element Name
Appropriate detector	Object	Vehicle Detector
Waiting traffic	State	Vehicle Detector
Turn or pedestrian light	Object	Vehicle Light Assembly Pedestrian Light Assembly
Durations	Parameter for timeout transitions	Various statecharts
Light times	Parameter for timeout transitions	Various statecharts
Appropriate parameters	Attributes	System configuration object
Table	UN	
Values	UN	
Defaults	UN	
Mode 2 Parameters	Class	System configuration class
Reset parameters	Operation	System configuration class
Parameter	Attribute	System configuration class
Value type	UN	
Description	UN	
Primary green time	Attribute	System configuration class
xx to yyy seconds	Attributes	System configuration class
Length of time	UN	
Primary green light	Object	Instance of Vehicle Light
Primary Yellow Time	Attribute	
Primary yellow light	Object	Instance of Vehicle Light
Primary Red Delay Time	Attribute	
Primary red light	Object	Instance of Vehicle Light
Primary walk time	Attribute	
Secondary green light	Object	Instance of Vehicle Light
Primary WALK light	Object	Instance of Pedestrian Light
Primary warn time	Attribute	
Primary FLASHING DON'T WALK light	State	Of an instance of Pedestrian Light
WALK light	Object	Instance of Pedestrian Light

Table 10.4 Roadrunner fixed cycle mode nouns (continued)

Noun Phrase	Element Type	Element Name
Primary turn Green time		
Primary turn light	Object	Instance of Vehicle Light
Primary turn light parameter	Attribute	Of an instance of Vehicle Light
Primary turn yellow time	Attribute	Of an instance of Vehicle Light
Default values	Attributes	Various classes
System configuration	Class	System configuration class
Default cycle times	Attributes	System configuration class
Turn lane	Object	Instance of Vehicle Light Assembly
Ped signal	Object	Instance of Pedestrian Light Assembly
Values	UN	
Table 3	UN	
Direction	UN	
Primary road	Object	Instance of Vehicle Light Assembly
Car	Actor	Vehicle
Turn lane	Object	Instance of Vehicle Light Assembly
Pedestrian	Octor	Pedestrian
Secondary road	Object	Instance of Vehicle Light Assembly
Timing diagram	UN	
Cycle times	Attributes	Various statecharts
Simultaneous turn lane mode	State	System Object
Road	UN	
Straight traffic	Object	Instance of Vehicle Light Assembly

The causal agent strategy for the Fixed Cycle Mode identifies the elements that cause things to change: these are time (that is, the internal elements themselves decide when to change the light based on time; these elements include the Vehicle-LightAssemblies and PedestrianLightAssemblies), vehicles (in the turn lane only in this case) and pedestrians.

Two views of the object analysis are useful here. First is to recap the system architecture view because most of these elements are required for fixed cycle mode. This is shown in Figure 10.3. Since this is a structure diagram, it shows the system as a whole with its parts (object roles). However, we'd also like to see how the attributes are allocated to the classes, so that aspect is shown in Figure 10.4. Remember that the parts in the former figure are specified by classes, and in that figure there is more than one instance role specified by the same class (for example, P_VehicleLightAssembly and S_VehicleLightAssembly are both instance roles of the class VehicleLightAssembly—one for the primary road and one for the secondary road). Thus we could show the attributes in both or either but I think it just confuses the issue. I find it clearer to show the structure on one diagram and the class specifications on another.

By the way, note that in fixed cycle mode, we still need to have the sensors present for the vehicle turn lanes and for the pedestrian, although not for the through traffic. The different attributes for green, yellow, red, walking and warning times for the secondary and primary roads (and both through and turn traffic lanes) can be set in the appropriate instances of the VehicleLightAssembly and PedestrianLightAssembly. That is, if the instance for the vehicle light is to control the primary turn lane, then its greenTime attribute will hold the PrimaryGreenTurnTime property specified for the system. The instance for the secondary through light also has a greenTime attribute but it will hold the system property SecondaryGreenTime. At this point, we are assuming that the turn light controls can be all structurally and behaviorally identical but their role will be known to the IntersectionController. Later, we can validate this notion when we add the state behavior and execute the model.

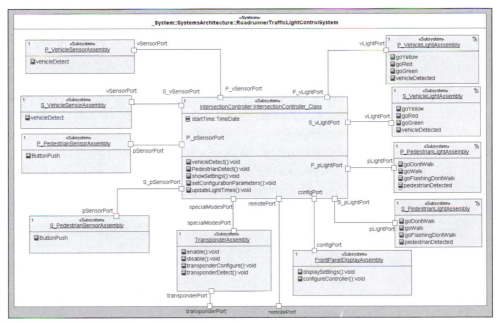

Figure 10.3 Fixed cycle mode collaboration Strategy 2 overall model

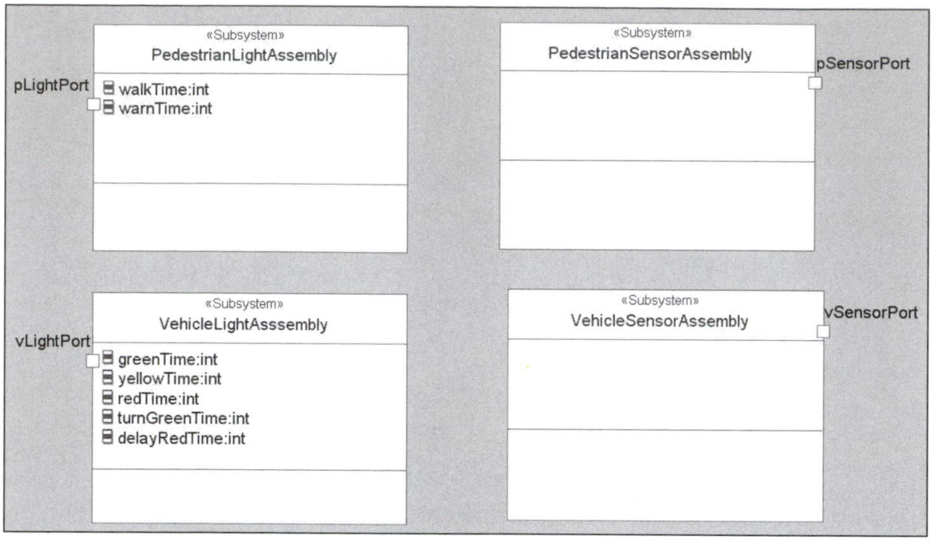

Figure 10.4 Fixed cycle mode collaboration Strategy 2 attributes

The UAV problem is far larger, but the strategy works because the object analysis is performed on use cases of the subsystems rather than the system as a whole. However, it can still be a complex task to model each subsystem-level use case for a complicated system such as the Coyote UAV. In the sample problem statement below, we have only underlined the nouns relevant to the Acquire Image use case.

The Unmanned Air Vehicle (UAV)

The Coyote UAV is meant to be a multipurpose reusable UAV with multimission capability. It is meant to operate at altitude of up to 30,000 feet with ground speeds of up to 100 knots (cruise) and 150 knots (dash) and carry a payload of up to 450 lbs for durations in excess of 24 hours. The Coyote is meant to fly unimpeded in low-visibility environments while carrying either reconnaissance or attack payloads. While controllable from the ground station CMPCS, it is also capable of flying complex flight plans with specific operational goals of systematic area search, ground route (road-based) search, and orbit surveillance of point targets. Coupled with manned control from the ground, the Coyote provides sustained 24-hour flight with real-time visual, infrared or radar telemetry, with target recognition preprocessing. Communications are jam-resistant, although need not be antijamming in a high ECM environment. Control commands shall be encrypted while telemetry data can be compressed but unprotected. Telemetry rates for visual telemetry support 30 frames-per-second (fps) at 640 × 400 resolution. Range of flight is meant to be fully supported within <u>line of sight (LOS) range</u> but since the <u>Coyote</u> also has the ability to be passed among different <u>CMPCSs</u>, its range is considerably greater than LOS. For navigation, the Coyote has on-board Global Positioning System (GPS) based-navigation as well as being directly controllable from the ground station.

Unlike many smaller UAVs, the Coyote does not require specialized launch and recovery vehicles. It can use a short runway for either automated or remote-controlled takeoff and landing.

The Coyote Mission Planning and Control System (CMPCS)

Mobile CMPCS with capability to control up to four UAVs with a manned control station per UAV that fits into a smaller towable trailer. Each control station consists of two manned substations—one for controlling the CUAV and one for monitoring and controlling payloads. If desired, both functions can be slaved together into a single control substation. Control of the aircraft shall consist of transferring navigational commands which may be simple (set altitude, speed, direction), operational (fly to coordinate set, orbit point, execute search pattern, etc.), planned (upload multisegment flight plan) or remote controlled with a joystick interface. Stable flight mechanics shall be managed by the aircraft itself but this can be disabled for remotely controlled flight.

The CMPCS displays real-time reconnaissance data as well as maintaining continuous recording and replay capability for up to 96 hours of operation for four separate CUAVs. In addition, with attack payloads, the Coyote can carry up to four Hellfire missiles with fire-and-forget navigation systems.

The Unmanned Air Vehicle (UAV)

...

Mission Modes

Beyond flight modes, CUAV shall be designed for highly flexible mission parameters. Normal mission modes include:

- Preplanned reconnaissance
- Remote controlled reconnaissance
- Area search
- Route search
- Orbit point target
- Attack

A mission can consist of any number of sequential submissions each operating in a different mission mode, depending on the current payload.

The Coyote Mission Planning and Control System (CMPCS)

The CMPCS is housed in a 30x8x8 triple-axis trailer that contains stations for pilot and payload operations, mission planning, data exploitation, communications, and SAR viewing. The CMPCS connects to multiple directional antennae for communication with the CUAVs. All mission data is recorded at the CMPCS since the CUAV has no on-board recording capability. The CMPCS has a UPS that can operate at full load for up to 4 hours in addition to using commercial power or power generators.

A single CMPCS can control up to four CUAVs in flight with one station per CUAV. Each CUAV control station provides both pilot and payload operations with separate control substations, although both functions can be slaved to a single substation for low-vigilance use.

For the reconnaissance payloads, the CMPCS shall provide enhanced automated target recognition (ATR) capability for all surveillance types—optical, infrared, and radar. While the CUAV has a rudimentary capability, the CMPCS provides much more complete support for the quick identification of high value targets in the battlefield. This capability is specifically design to identify mobile and time-limited targets that may only be exposed for brief periods of time before they go back into hiding. The system is expected to provide high clutter rejection and a low false-positive error rate. The ATR shall be able to identify and track up to 20 targets within the surveillance area, with likely identification and probability assessments for each. In addition to the ATR, the payload operator can add targets visually identified from reconnaissance data or gathered from other sources. The battlefield view can be transmitted over links to remote command staff for tactical and strategic assessment.

Image Acquisition and Processing

Images may be acquired from all three sensor platforms—optical, FLIR (forward looking infrared), and SAR (synthetic aperture radar). Optical and FLIR are passive systems using emitted energy from terrain or targets to gather information. The SAR is an active sensor, painting relatively stationary targets with energy (in the microwave range) and using the reflection of these pulses to determine reflectivity and altitude. The optical and FLIR resolution for single images may be as high as 1900x1600 resolution while real-time video for all sensor platforms is limited to 640x480 resolution at a rate of 30 fps. Streaming imagery shall be sent with enough redundancy so that complete loss of random frames shall not affect the quality of other frames. The sensor platforms may be focused at any range from 10 meters to infinite and may be zoomed up to 100x actual. The sensor platforms shall be mounted on a gimbaled assembly so that the system can be aimed without affecting the attitude of the CUAV. The FLIR includes a laser range finder to determine target range so that it can be used in fire control applications. The SAR shall emit a series of pulses meant to emulate the behavior of much larger physical aperture antenna; the images from the SAR are a combination of timed radar surface reflections to be combined into a SAR image in the ground station using Fourier transforms. Thus, a single SAR image results from a set of images each resulting from a single radar pulse from the SAR platform, but combined in the ground station. Aiming the SAR is done through the use of Doppler sharpening, limiting the amount of information that must be transmitted to the ground station to construct the SAR image. The use of two pulse emitters in the SAR platforms allows the interference patterns to be constructed providing altitude determination as well as radar reflectivity data.

Images may be compressed using lossy or nonlossy methods to minimize communication bandwidth requirements. The JPEG 2000 compression standard shall be used; for streaming video the associated MJP2 standard shall be used The compression may be set dynamically by the payload operator to be 0% to 80% with the default setting to be nonlossy 50% compression. The imaging system is required to achieve the desired compression only within 20% of the requested due to the variances in the image contents. The selection of lossy or nonlossy compression shall be determined automatically by the imaging system, switching to lossy compression only when the desired compression rate cannot be achieved using lossless compression.

Table 10.5 shows the underlined noun phrases in the problem statement. We have tried to only underline and include noun phrases that are related to image acquisition and processing. However, this is one of the difficulties in applying this strategy in large-scale systems. Unless the textual description/specification is extraordinarily well-organized, it is difficult to just underline the noun phrases for the use case of immediate concern, and the underlining of every noun phrase leads to hundreds or thousands of candidate classes that must be sorted and organized.

Table 10.5 CUAV Acquire Image nouns

Noun Phrase	Element Type	Element Name
Real-time visual, infrared or radar telemetry	Class	Image
Target recognition preprocessing	Algorithm	PreprocessImage()
Communications	UN	
Control commands	Class	Command
Telemetry data	Class	Image
Telemetry rates	Constraint	Telemetry Rates
Visual telemetry	Class	VisualImage
Line of sight (LOS) range	UN	
Coyote	Class	CUAV
CMPCSs	Class	GroundStation
Reconnaissance payloads	Class	ReconAssembly
Enhanced automated target recognition (ATR) capability	Algorithm class	identifyTarget() Target
Optical	Class	VideoAssembly
Infrared	Class	FLIR_Assembly
Radar	Class	Radar_Assembly
High value targets	Class	Target
Battlefield	UN	
Capability	UN	
Mobile and time-limited targets	Class	Target
System	Class	CUAV_Project
False-positive error rate	UN	
ATR	Algorithm class	identifyTarget() Target
Targets	Class	Target
Surveillance area	UN	
Payload operator	Actor	MissionSpecialist
Targets	Class	Target
Reconnaissance data	Class	Image
Battlefield view	Class	Image
Links	UN	
Images	Class	Image
Sensor platforms	Class	ReconAssembly

Table 10.5 CUAV Acquire Image nouns (continued)

Noun Phrase	Element Type	Element Name
Optical	Class	Video_Assembly
FLIR	Class	FLIR_Assembly
Passive systems	UN	
Emitted energy	UN	
Terrain	Class	TerrainFeature
Targets	Class	Target
Information	Class	Image
SAR	Class	Radar_Assembly
Active sensor	UN	
Targets	Class	Target
Energy	UN	
Microwave range	Attribute	Frequency
Reflection	UN	
Pulses	Attribute	PulseRate PulseLength
Reflectivity	Attribute	Color
Altitude	Attribute	Range
Optical and FLIR resolution	Attribute	resolution
Real-time video	Class	ImageStream
Sensor platforms	Class	ReconAssembly
Resolution	Attribute	X_resolution Y_resolution
Rate	Attribute	FrameRate
Streaming imagery	Class	ImageStream
Redundancy	UN	
Frames	Class	Image
Quality	UN	
Range	Attribute	Range
Gimbaled assembly	Class	Gimbal
Attitude	UN	
CUAV	Class	CUAV
Laser range finder	Class	LaserRangeFinder
Target range	Attribute	Range
Fire control applications	UN	

Table 10.5 CUAV Acquire Image nouns (continued)

Noun Phrase	Element Type	Element Name
SAR	Class	Radar_Assembly
Series of pulses	Class	Pulse
Physical aperture antenna	UN	
Images	Class	Image
Combination of timed radar surface reflections	Class	Radar_Image
SAR image	Class	Radar_Image
Ground station	Class	GroundStation
Fourier transforms	Algorithm	ComputeSARImage()
Single SAR Image	Class	Radar_Image
Set of images	Class	Radar_Image
Single radar pulse	Class	Pulse
SAR Platform	Class	Radar_Assembly
Information	UN	
Pulse emitters	Class	PulseEmitter
Interference patterns	Class	Radar_image
Altitude determination	Attribute	Altitude
Radar reflectivity data	Attribute	Color
Lossy or nonlossy methods	Algorithm	CompressImage()
Communication bandwidth requirements	UN	
JPEG 2000 compression standard	UN	
MJP2 standard	UN	
Compression	Algorithm	CompressImage()
Payload operator	Actor	PayloadOperator
Default setting	Attribute	DefaultCompression
Imaging system	Class	CUAV
Desired compression rate	Attribute	CompressionRate
Variances in the image contents	UN	
Selection	Attribute	IsLossy
Lossless compression	Attribute	IsLossy

In this problem, the use case Acquire Image is a capability of the Reconnaissance Management subsystem. A part of this analysis might spill over into other subsystems (notably Targeting, Airborne Datalink, and the reconnaissance payloads), but primarily we'll be focused on two things:

1. Identifying and linking together the objects, classes, attributes, and algorithms within the Reconnaissance Management subsystem

2. Characterizing the interfaces of the Reconnaissance Management subsystem to its peer subsystems

To that end, it is necessary to decide how to split the functionality between the payloads and the other subsystems. In general, the functional allocation will be thus:

- The recon platforms are responsible for the production of a single uncompressed and unprocessed image at a time.

- The reconnaissance management subsystem (RMS) is responsible for processing and compressing recon images, managing the streaming of images, commanding, aiming, and focusing the recon platforms, and to identify targets (as directed by the Targeting subsystem).

- The RMS is responsible for commanding configuration of the recon platforms.

- The Targeting subsystem is responsible for target selection, setting of the target identification parameters, and tracking targets.

The created class diagram in shown in Figure 10.5. You'll note that some of the identified objects in the previous table are not represented in the class diagram. This is because they are allocated to other subsystems; for example, the Gimbal is a part of the payload, not part of the Reconnaissance Management subsystem. Also, there are a few classes in the diagram that do not appear in the table. They were added to hold operations or attributes, or because it seemed likely that it would be needed based on the other classes and how they must interact. The model is certainly not complete, but this is typical when such a strategy is applied. Multiple strategies must be applied to find all the classes and their features to fully realize a use case.

The second strategy, "Causal Agents," looks for objects that initiate actions. Many of the actions are ultimately caused by the Payload Operator, such as moving the gimbal, aiming and focusing, setting the resolution and so on. There must be some facility in the subsystem that accepts and processes those commands. Once initialized, the recon platforms themselves produce images that they send to the Reconnaissance Management subsystem. Lastly, the Targeting subsystem specifies the nature of the targets to look for and the reconnaissance management subsystem must receive and respond to such commands as well. Figure 10.6 shows the classes

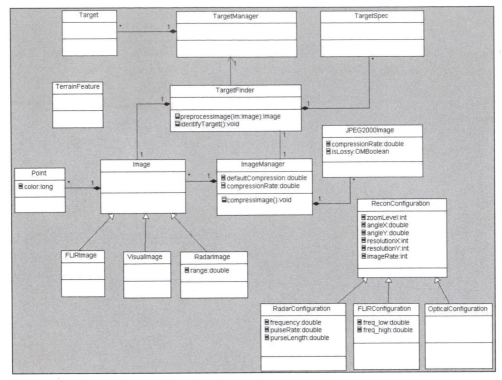

Figure 10.5 Acquire Image collaboration Strategy 1

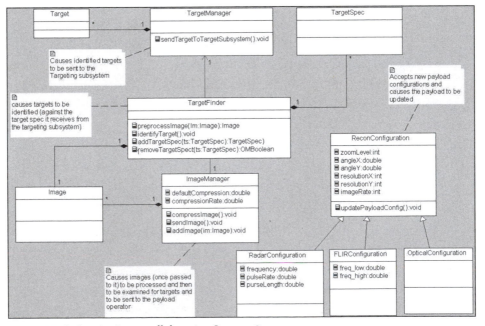

Figure 10.6 Acquire Image collaboration Strategy 2

identified using the casual agent strategy. Some operations were added to facilitate the roles of the classes in causing the appropriate actions to occur.

Answer 5.2   Apply Services and Messages Strategies

These strategies are all about using the interfaces between the system and its actors or between elements inside the system to identify objects. In the case of a large-scale and abstract service, the service must be decomposed into subservices, and possibly sub-subservices. Ultimately, whether or not services are decomposed, the identified services must be provided by the objects within the system. The purpose of this strategy is to use this fact to identify objects inside the system and then allocate the services to those objects. Since services have both a (at least one) caller and a provider, the strategy will also identify links between the objects because, in the majority of cases, to invoke a service on a server object requires an association from a client object.

The strategies work in a different way as well. First of all, most of these services will require data to be passed, either as input, output, or bidirectional parameters, or as return values. Those parameters must come from somewhere and many of them are likely to be instances of classes or application-specific data types (which also need identification).

In the first half of this section, we'll look at the same two Roadrunner use cases—"Detect Vehicle" and "Fixed Cycle Time Mode." In the first case, clearly the messages have to do with a vehicle arriving. The specification states that there are three ways to detect vehicles: passive loop induction, laser range finding, and changes in the infrared field. While it is possible to maintain the distinction between how the detection occurred, we may assume that for the purpose of detecting an oncoming vehicle, how the detection occurs isn't as important as that it occurred and where (as in which traffic lane) it occurred. Thus, we probably have a service such as

vehicleDetected(d:Direction)

Where the type Direction may be an enumerated type with values of (PRIMARY, PRIMARY_TURN, SECONDARY, and SECONDARY_TURN) in this case.[2, 3] The two considerations for this service (in terms of using it as a strategy to identify objects) is who knows that a vehicle is detected and can therefore invoke the

---

[2]   This enumerated type obviously wouldn't work around the Arc de Triomphe, but it will for our problem case.

[3]   There is another way to detect vehicles using transponders for emergency and priority vehicles; however, there are other use cases that deal with that functionality.

services, and who cares that a vehicle has been detected and therefore provides the service. Obviously some (or possibly a set) of objects are keeping track of whether or not we should turn lights green to permit traffic to flow, depending, of course, on operational mode. The traffic lights control the traffic and so maybe they should be sent the message—but which one? All of them? Or should all the traffic lights be very simple and a centralized controller do all the work?

There are many valid answers to that question. In general, I prefer to apply the concept of "distributed intelligence" rather than "centralized intelligence." In my experience, it is far easier to create a set of many half-witted objects that collaborate than it is to create a single brilliant object in charge of everything, that controls a set of moronic slaves.[4] In this case, we'll take the approach that each traffic light assembly is responsible for managing its own vehicular traffic. In that case, each instance of Traffic Light Assembly will receive the messages from only its vehicular sensors (that is, the sensors for its through and turn lanes). This further implies that the traffic light assemblies must collaborate since, if the vehicle arrives in the secondary through lane, the secondary traffic light must notify the primary to go red so that the secondary traffic can proceed. This implies another service "Please Go Red" that one traffic light may send another.

Following this logic, we arrive at a class diagram similar to Figure 10.7.

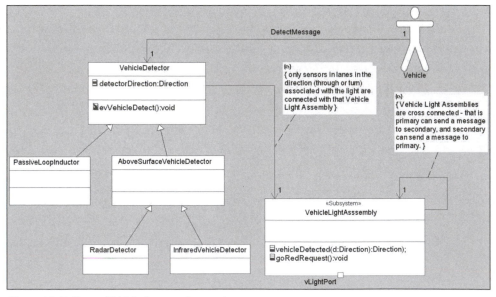

Figure 10.7 Detect Vehicle Strategy 3—services

---

4 I'm sure there's a political conclusion to be drawn there somewhere but I'll leave that as an exercise to the reader ;-)

The other use case under consideration is fixed cycle mode. In this mode, the behavior of the vehicle light assemblies is driven by time-based events. However, they do need to send messages to each other that inform one vehicle light assembly that it is safe for it to allow vehicles to pass. In addition, there must be some communication between the vehicle light assembly and the pedestrian light assembly since they must work in coordination. This strategy doesn't require communications from the vehicles, but it does require it from the pedestrian sensor assemblies.

The services from the pedestrian sensor assembly are similar to the vehicular case:

pedestrianDetected(d:Direction);

The services going back and forth between the pedestrian and vehicular light assemblies are a bit more complex, because these objects must clearly work in concert. In order to understand exactly how they interact, we must explicitly define how the objects work in isolation and what information they need in order to perform that behavior. The system-level behavior associated with this use case was given in the earlier discussion when we specified the fixed cycle mode use case. In this case, we are now going to allocate portions of this behavior to different objects; each vehicular light assembly will be responsible for controlling traffic in only one direction, although this includes both through and turn lanes, as well as coordinating with the pedestrian light assembly in the same direction, and notifying the vehicle light assembly in the orthogonal direction when it is safe to allow traffic to flow. The pedestrian light shall only be concerned with managing pedestrian traffic.

The following two figures show roughly what might be reasonably expected for the state machines for these classes. However, the state machines are not complete (we'll do that later). The purpose of doing them now is to help us understand how the two objects collaborate. The simpler of the two is the pedestrian light assembly, in Figure 10.8.

The vehicle light assembly state machine is a bit more complex (see Figure 10.9). It uses one and-state to control the interaction with the pedestrian light assembly, another to manage the turn light, a third to keep track of waiting vehicles in the turn lane, and another for management of the through traffic light.

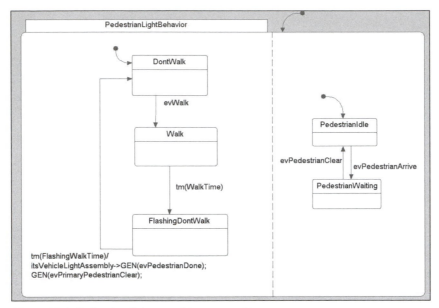

Figure 10.8 Pedestrian light assembly state machine

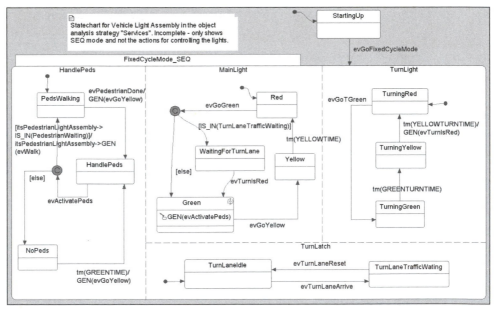

Figure 10.9 Vehicle light assembly state machine

Thus, we can see that the pedestrian light sends an evPedestrianDone event to the vehicle light assembly, while the vehicle light assembly sends an evWalk event to the pedestrian light assembly.[5] And, of course, something has to provide the services to actually turn on the lights for the vehicles and pedestrians.

These considerations allow us to refine our object model a bit more, as shown in Figure 10.10.[6]

Let us now turn our attention to the UAV. There are many services associated with Acquire Image. Of course, different payloads are involved, but they are outside the scope of the subsystem we're modeling here. Nevertheless, they must provide services to aim and focus the sensor systems and provide the image. Let us assume that while the reconnaissance management subsystem may initiate the creation of an image, that image is produced asynchronously. Furthermore, the recon payloads may be set up to send a stream of images at some specified rate. This means that the reconnaissance management subsystem requires a service to receive those images. The image itself is surely an instance of a class and may contain metadata such as

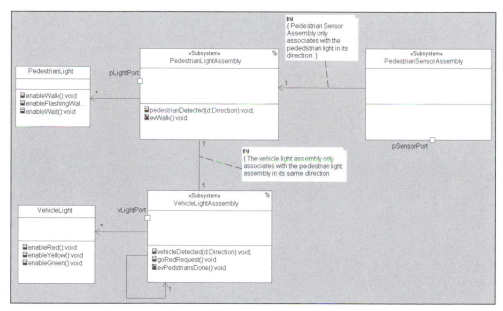

Figure 10.10 Fixed cycle mode Strategy 3—services

---

[5] The IS_IN operator, which allows an object to examine another's current state, is provided by the infrastructure for free so it won't be explicitly modeled.

[6] Only the objects identified in this strategy are shown in this diagram in an effort to be clear about which objects were identified using the strategy. In the actual project, you would just add the object, classes, etc. to the existing collaboration rather than create a separate diagram.

the location (via the navigation subsystem GPS), date and time stamps, the kind of sensor used to generate the image, configuration parameters of the sensors (such as the wavelengths used, or pulse width and frequency from the SAR), the encryption method, etc.

A first-cut solution with the services strategy is shown in Figure 10.11. The only use of the ports in the collaboration is to connect to the ports of the encapsulating Reconnaissance Management subsystem. The provided ports are "behavioral" meaning that they provide the services requested over that port. The required interfaces mean that that is the class whose instance will ultimately invoke those services from the other subsystems over those ports. The ports between subsystems are typically characterized by both provided and required interfaces. In this collaboration, the provided and required interfaces are typically realized by different classes inside the Reconnaissance Management subsystem. We've added the interface names to clarify the port usage.

The collaboration is a bit elaborated from the previous strategy. This elaboration includes moving operations and attributes around and creating new classes entirely. This process of elaborations is called "refactoring" and is very common during the creation of an object analysis model.

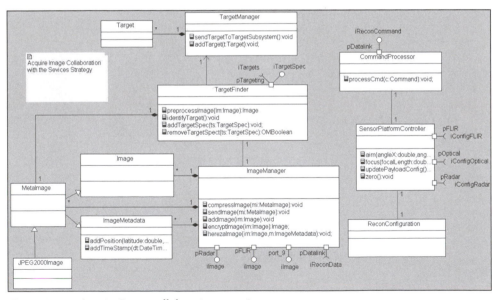

Figure 10.11 Acquire Image collaboration—services strategy

Answer 5.3 Applying the Real-World Items and Physical Devices Strategies

In the traffic light control system, there are both real-world items and, of course, a number of physical devices. In terms of the real-world resources to be managed, we need to perform vehicle counting and compute traffic flow statistics, and in the adaptive mode, the timing of the traffic lights change to optimize the traffic flow. Thus "cars" are a resource that must be represented and used in computational analysis for the Detect Vehicle use case (which includes management of traffic statistics). As for physical devices, there are the vehicle sensors—passive loop inductor, infrared vehicle detector, and radar vehicle detector, the pedestrian sensor ("button"), and many different actuators (lights to be turned on and off) in the correct sequences and timing.

The classes shown in Figure 10.12 representing the sensors and actuators do not simulate their behavior but rather provide interfaces to control and query the actual physical devices. The vehicle is represented as a summary statistic for the vehicle traffic flow. The application of this strategy resulted in the identification of a new class (VehicleStats) and also a new operation for the Intersection Controller. In

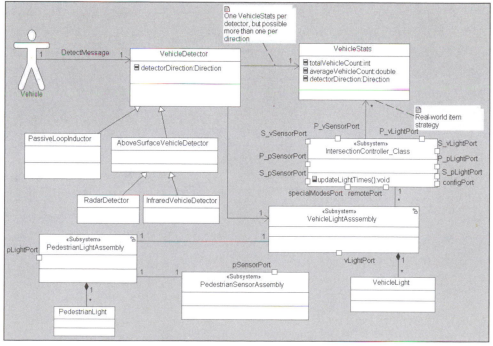

Figure 10.12 Detect Vehicle use case—real-world items strategy

addition, physical devices (VehicleLight and PedestrianLight) have also been added. The physical devices for sensors have already been identified but are not any more detailed in this model.

The Fixed Cycle Time mode is primarily concerned with the actuators because the behavior of this use case is time-based, rather than arrival-event based. However, it doesn't add any new classes over those in the previous figure so we won't redraw it.

The Coyote UAV also has many real-world items and physical devices for the Reconnaissance Management subsystem. In terms of resources, Targets are certainly real-world items that must be identified and (by the Targeting subsystem) tracked. Images are a representation of "ground truth" and so represent the real-world condition. As far as physical devices, the recon platforms configurations are managed by the Reconnaissance Management subsystem, and so fit the strategy parameters as well. The classes identified by this strategy are shown in Figure 10.13.

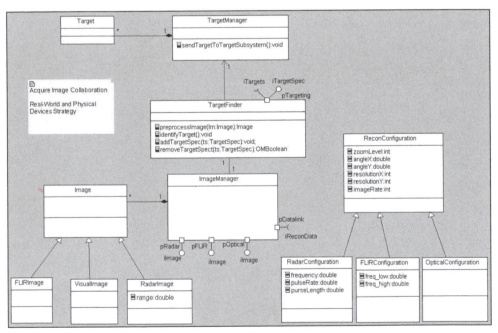

Figure 10.13 Acquire Image use case—real-world items strategy

Answer 5.4   Apply Key Concepts and Transaction Strategies

The Roadrunner traffic light control system is fairly simple in terms of key concepts. A short list of concepts is shown in Table 10.6. Note that this list only provides the concepts around detecting vehicles because we are analyzing the use cases independently. A different list would be constructed for a fixed cycle mode or other use case. In addition, some of the concepts listed overlap (such as Vehicle and Vehicle Detection) and may have the same realization within the model.

Table 10.6 Roadrunner Detect Vehicle key concepts

Concept	Description	Representation
Lane	The position of a car is important in terms of which lane it is in, so that it can be controlled via the light.	Class: Vehicle Detector (one instance per lane) Attribute: detectorDirection
Time	Time is fundamental for time-based state transitioning for light control	Attribute: many attributes for times for different lights Event: many timeout events in statecharts
Traffic	The flow of cars along a specific road or through a specific intersection	Actor: Vehicle Event: evVehicleDetect Attribute: totalVehicleCount, averageVehicleCount
Vehicle	A car traveling on one of the lanes monitored and controlled by the system	Actor; Vehicle Event: evVehicleDetect
Vehicle Count	The number of vehicles traversing the lane during a period of time	Attribute: totalVehicleCount, averageVehicleCount
Vehicle Detection	A detection of a car in the lane	Actor: Vehicle Event: evVehicleDetect

As for the transaction strategy, the arrival of a pedestrian or vehicle can be considered a transaction. Remember that a transaction represents an interaction between objects. In general, the transaction includes the request but isn't complete until the response initiated by that request has completed.

The "Request for permission for traffic to flow" is represented in the statechart for the VehicleLightAssembly where we have the "latch" and-state remembering that a vehicle arrived for the various control use case. In the fixed cycle mode use case, we only need to remember the turn lane (because the through lanes are driven by time) and the pedestrians. In other modes of operation, we will need to remember when vehicles arrive for other lanes as well. The latch is cleared when the request is handled by sending an evTurnLaneReset once the event is handled. Also, note that in most transaction-oriented systems, each request results in a new transaction that must be managed. In this case, many requests (multiple cars or anxious pedestrians) may result in only a single transaction since the latch and-state throws away additional requests once a single one is made.

Combining these two strategies together results in Figure 10.14.

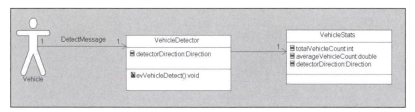

Figure 10.14 Detect Vehicle use case—key concepts strategy

When there are multiple transactions to be managed, each is normally represented as a separate object. Since there is only one in this case, we integrated it into the statechart. However, we could have done this by creating a WaitingCar class with a multiplicity of 0,1 to the VehicleDetector class. The VehicleDetector could dynamically create the instance when a car arrives and the VehicleLightAssembly could destroy it when the light cycles through, handling the waiting car transaction. I recommend you create this diagram as an exercise to the reader.

The CUAV Acquire Image use case has abstract concepts as well, around what an image is, its acquisition, and possibly early processing. These concepts are shown in Table 10.7 with brief descriptions and how they are represented in the model.

Table 10.7 CUAV Acquire Image use case key concepts

Concept	Description	Representation
Point	Primitive element of a picture that holds a color	Class: Point
Image	Picture of something of interest	Class: Image
FLIR Image	Image taken by a forward-looking infra-red system, colored by translation into "false color"	Class: FLIRImage
Visual Image	Image taken by an optical camera in the human visual range	Class: VisualImage

Table 10.7 CUAV Acquire Image use case key concepts (continued)

Concept	Description	Representation
Radar Image	Image taken by painting an area with active radar frequencies and capturing the reflection	Class: RadarImage
Focus	The focal length of the camera; also the ability to adjust the focal length of the camera	Attribute: focalLength
Zoom	The ability to constrain an image to a smaller area while simultaneously increasing the resolution of that selected area	Attribute: zoomLevel
Aim	The ability to direct the imaging system to look at different areas	Attribute: angleX, angleY
Image Resolution	The degree of detail held within an image	Attribute: resolutionX, resolutionY
Compression	The ability to reduce the memory required to hold an image	Class: JPEG2000Image Attribute: compressionRate Operation: compressImage()
Lossy Compression	Compression that results in a loss of detail	Attribute: isLossy
Lossless Compression	Compression that does not result in a loss of detail	Attribute: isLossy
SAR Frequency	Radar frequency used to paint the target area with a synthetic aperture radar	Attribute: frequency
SAR Pulse Rate	The rate at which radar pulses are delivered to paint a target area	Attribute: pulseRate
SAR Pulse Length	The duration of a radar pulse	Attribute: pulseLength
FLIR Frequency Range	The range of frequencies examined by an infrared camera	Attribute: freq_low, freq_high

As far as transactions go, the ground station can request images to be delivered either individually or by requesting images to be taken at a specified rate. The first isn't what we normally mean by a transaction because the lifetime for the transaction is just until the (synchronous) request is handled. Transactions, as a strategy of identifying objects, have a nontrivial lifetime and the request is typically asynchronous.

Requesting the delivery of a set of messages at a certain rate is closer to the notion of a transaction than a simple request for an image. Because the images need not be stored within the UAV (they are actively transmitted but not necessarily stored within the UAV), but delivered as soon as they are ready, the representation of this transaction is simply to have a nonzero image rate. To stop the transmission of a set of images, the mission specialist need only set this attribute to zero.

The classes and class features identified using these strategies results in Figure 10.15.

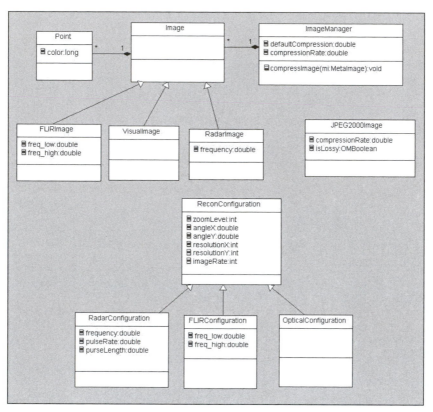

Figure 10.15 Acquire Image use case key strategies and transactions strategy

Answer 5.5 Identify Visual Elements and Scenarios Strategies

This problem asks for the application of the "Apply Scenarios" strategy for object identification. As with all other object-identification strategies, this one is best applied in conjunction with one or more others. Nevertheless, it is a remarkably useful strategy to have in your analysis toolbox.

The first part of the answer is the application of this strategy to the Roadrunner Traffic Light Control System "Detect Vehicle" use case. The most obvious way to apply the strategy is to simply start at the first message in the use-case scenario, find the objects that deal with it and perform the necessary actions, and then go on to the next. Sometimes, you'll feel more comfortable starting in the middle of the scenario, either because it is of significant importance or because it is an area that is more understood than the rest. Whichever ordering is used, the job isn't complete until all the messages to and from the system executing the user case are accounted for.

The solution here is shown as a set of diagrams. The first shows the elaborated sequence diagrams and the last depicts the resulting class diagram. Note that the display and keyboard classes are not as detailed as they could be. If the UI is under design as well, then the elements of the display (text, icon, and other graphical widgets) would be identified as a part of this effort along with their relevant features (operations and attributes).

The first set of sequence diagrams is shown in Figure 10.16 through Figure 10.18. In your own model, you might very reasonably choose to show all the messages on a single sequence diagram because you can easily set the zoom level and scroll as necessary. On the printed page, though, it is easier to follow if the single scenario is broken up a bit more. Since UML 2.0 provides a standard way to do that ("referenced interaction fragments"), that's what's been done for this solution to improve its readability on the printed page.

The problem statement back in Chapter 5 required this solution to be done by "elaboration" of the original sequence diagram; that is, take the original sequence diagram, copy it, and add lifelines and messages to show the same scenarios at a more detailed level of abstraction. To make this even clearer, I've emboldened the lifelines from the original scenario. Notice also that I've kept the "use case" lifeline, which is a stand-in for the "system executing the use case." This makes the comparison of the original and elaborated sequences much more straightforward.

Figure 10.16 shows the first level of elaboration. The scenario breaks up into two parts: the configuration of the passive loop inductor and the subsequent detection and counting of vehicles. Figure 10.17 elaborates that configuration into two parts, basic configuration and enabling the detector, and Figure 10.18 shows the details of the setting up the basic configuration. In these latter two figures, human

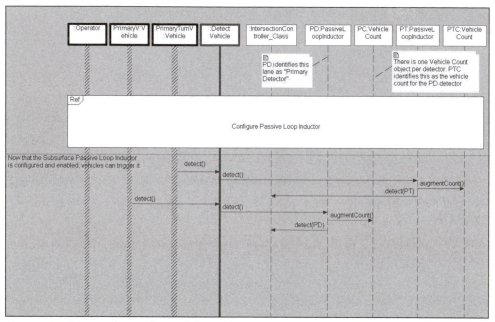

Figure 10.16 Passive loop inductor elaborated scenario

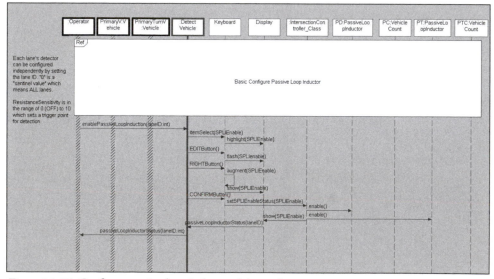

Figure 10.17 Configure passive loop inductor elaborated scenario

interface objects (Keyboard and Display) are introduced to show how the human operator interacts with the HMI[7] front panel to actually perform the configuration. As mentioned earlier, if the design of the HMI was a focus of this work, we would have identified more detailed interface objects, such as Knob, Button, Key, TextDisplayWidget, and so on. These have been elided here, but would probably show up in a more detailed application of the strategy.

The only reason for breaking up Figure 10.17 was readability—the scenario was a bit long. So I decided to break it at a natural breaking point, even though it wasn't halfway through. This breaking point was the completion of the configuration of the passive loop detector. Figure 10.17 then goes on to show how the detector is enabled. One thing that is apparent from the scenarios is how the user interacts with the front panel display to make the setting changes. This permits user workflow analysis to be done to see if the HMI adequately meets the customer needs.

Another thing that is apparent from the scenario is how the objects collaborate to achieve the higher-level behavior specified by the user case. Note that the counts are done on a per-detector basis, so that detailed traffic-flow analysis can be done, if

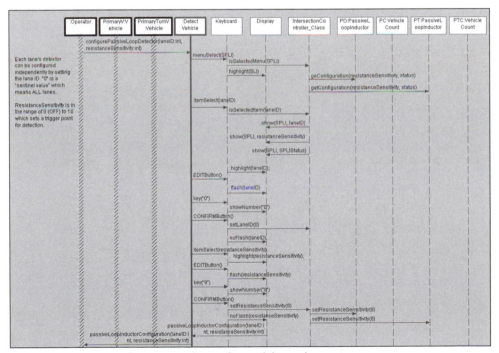

Figure 10.18 Basic configure passive loop inductor elaborated scenario

---

[7]   Human-machine interface

necessary. Also note that the detectors can be individually configured (more scenarios …) but in this case, we configured them all at once using the "0" sentinel value to select all passive loop detectors.[8]

Another thing to notice is the central role that the Intersection Controller plays in the interaction. Rather than have direct connections to every element to the front panel, the Intersection Controller mediates those connections. This was done to simplify the topology. The Intersection Controller clearly needs to have connections to the detectors anyway.

The next set of figures deals with the set-up and use of the infrared detector. This is also a passive detector, but its configuration is slightly more involved than the passive loop detector.

Figure 10.19 has pretty much the same structure as Figure 10.16, but with different internal parts to perform the infrared vehicle detection.

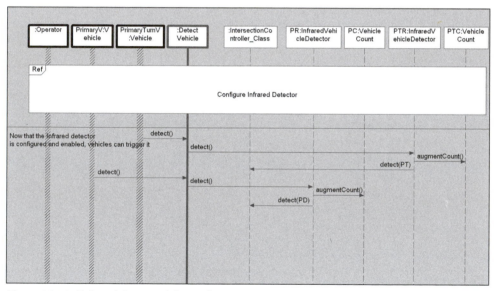

Figure 10.19 Infrared detector elaborated scenario

---

[8] You may be noticing the "scenario explosion" implied by referencing other, yet-to-be-created, scenarios. There are an infinite set of scenarios; however, it is adequate to create the minimal spanning set of scenarios, in which each requirement (or transition, if you have a state machine for the use case) is represented at least once.

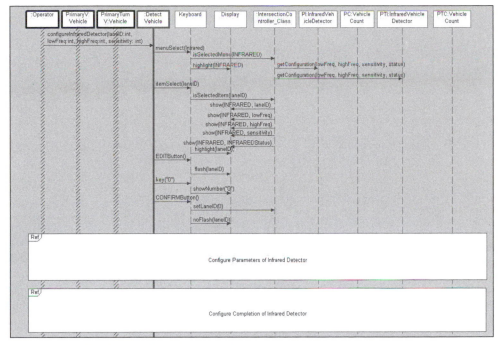

Figure 10.20 Configure infrared detector elaborated scenario

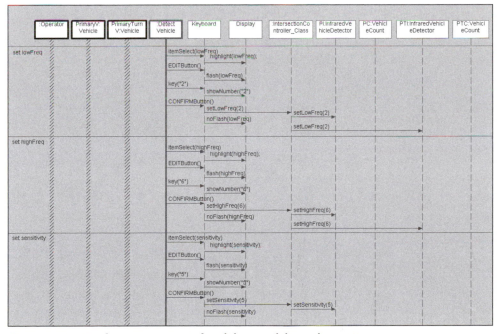

Figure 10.21 Configure parameters infrared detector elaborated scenario

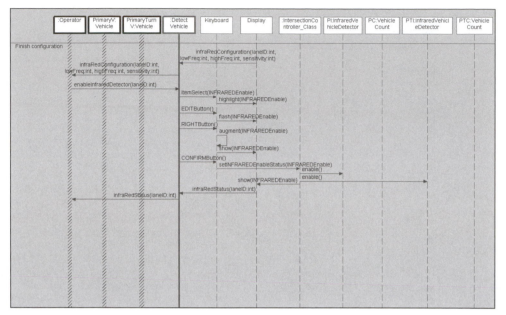

Figure 10.22 Complete configuration infrared detector elaborated scenario

The last scenario in this set is for the radar detector. The structure of these sequence diagrams is much the same as for the other scenarios. This scenario is broken up across three sequence diagrams as well, Figure 10.23 through Figure 10.25.

Figure 10.23 shows the highest-level sequence diagram, which includes the Configure Radar Detector reference interaction fragment. This interaction fragment is elaborated in Figure 10.24. Figure 10.24 in turn contains the reference interaction fragment Basic Configure Radar Detector that is subsequently displayed in Figure 10.25.

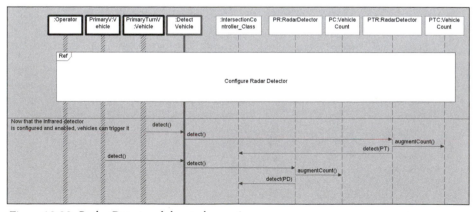

Figure 10.23 Radar Detector elaborated scenario

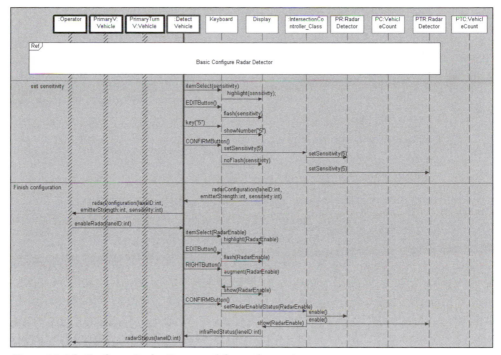

Figure 10.24 Configure Radar Detector elaborated scenario

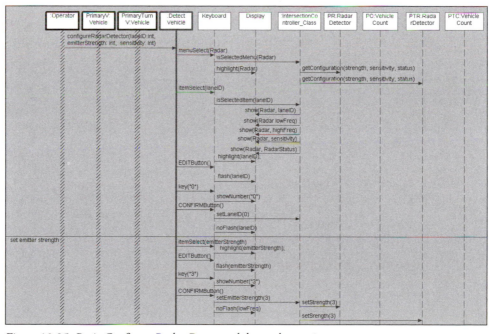

Figure 10.25 Basic Configure Radar Detector elaborated scenario

The last scenario presented here is the elaboration of the Get Stats use-case scenario. As before, this scenario is broken up into multiple sequence diagrams. The main one is shown in Figure 10.26. The first element in the sequence diagram is a reference to the second (Figure 10.27) that shows how the statistics are reset.

The original Get Stats use-case scenario contained an unordered interaction fragment. We can't just use that as-is because, while the order of the external events is unordered, the response to each event is not. Thus, we put the response to each detection event inside a strict interaction fragment, indicating that strict ordering is followed within that fragment. I also slightly modified the original scenario at the end, when statistics are requested. Specifically, I passed in specific lane numbers for the requests—"PD", a constant referencing the primary through lane (nominally equal to "1") and "0" to get the total for all lanes.

As before, Figure 10.26 contains a reference interaction fragment, Reset Stats, which is detailed in Figure 10.27.

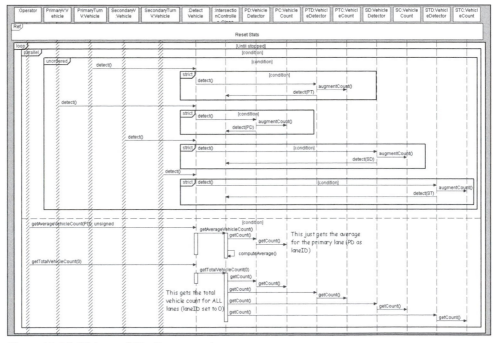

Figure 10.26 Elaborated Get Stats scenario

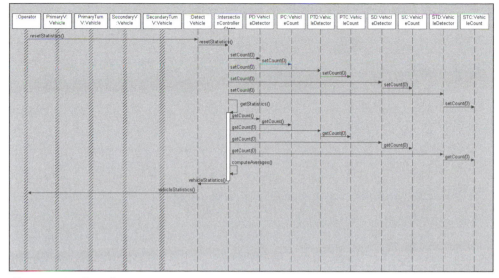

Figure 10.27 Reset Stats scenario

The resulting class diagram from this scenario analysis is shown in Figure 10.28.

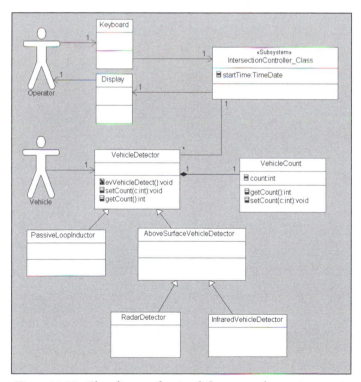

Figure 10.28 Class diagram for visual elements and scenario strategies

The second half of the problem is to apply the same strategies to the CUAV Acquire Image use case for the Reconnaissance Management subsystem. We could apply this strategy system-wide, in which case this would involve identification of objects within, at minimum, the airborne and ground data link subsystems, and the HMI for the missions operations specialist in the ground station in addition. In this solution, we'll limit ourselves to the Reconnaissance Management subsystem, which has no HMI.

In this solution, we're going to use the UML 2.0 lifeline decomposition feature. The original sequence diagram will be preserved (except for adding the explicit reference to the more detailed sequence diagram) and the detailed diagram will be added as a reference. Figure 10.29 shows the high-level sequence diagram from the user case. The next figure, Figure 10.30, shows the details as an elaboration of the referenced lifeline. Note that messages going into the Acquire Image lifeline on Figure 10.29 exit the ENV lifeline in Figure 10.30 and, conversely, messages coming from the Acquire Image lifeline in Figure 10.29 enter the ENV lifeline on Figure 10.30. However, notice that a few messages were "discovered" during the detailed analysis. This is common, and usually this means that the original use-case scenario should be updated to reflect this deeper understanding.

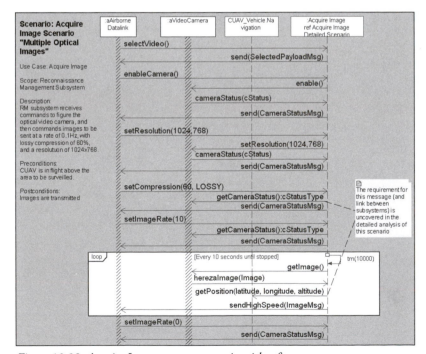

Figure 10.29 Acquire Image use-case scenario with reference

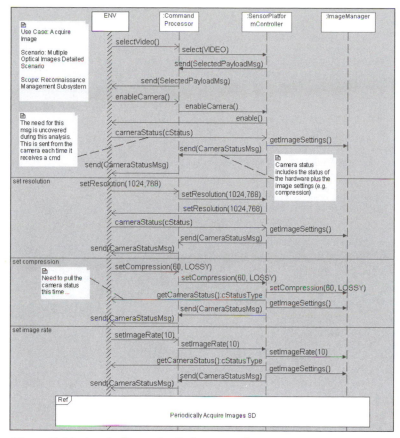

Figure 10.30 Acquire Image detailed sequence diagram

As before, to limit the length of the sequence diagrams, Figure 10.30 is decomposed using referenced interaction fragments. The referenced interaction fragment, shown in Figure 10.31, contains the message that handles the periodic acquisition and processing of the images for the scenario.

The result of this scenario analysis is the class diagram in Figure 10.32. The actors shown in the figure are other subsystems within the aircraft and are actors from the perspective of the Reconnaissance Management subsystem. To avoid name clashes with the actual subsystem class, such actors are identified with a leading "a" in the name, but really they are just stand-ins for the actual subsystem under development by another team.

For additional practice, try adding the HMI elements in the ground system to support the visualization and manipulation of the images, and add elements to support target identification and tracking.

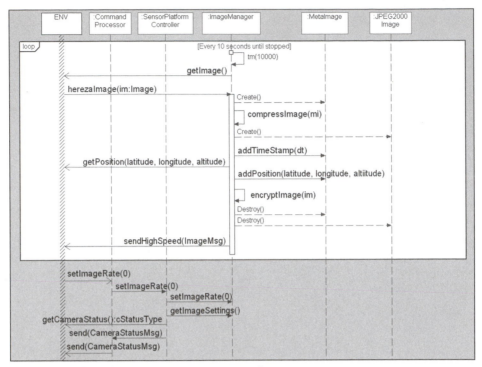

Figure 10.31 Periodic image acquisition sequence diagram

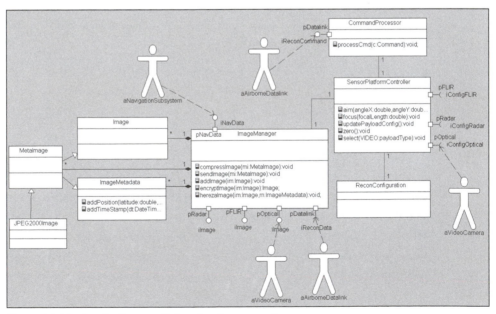

Figure 10.32 Acquire Image collaboration from scenario strategy

Answer 5.6 Merge Models from the Various Strategies

In this problem you were to merge the solutions from the previous strategies into a single class diagram to show the elements collaborating together. You should have noticed that, unsurprisingly, the strategies had significant overlap in the elements they identified, although each tended to find some unique elements as well. In practice, it is usually enough to use two to four strategies on any given use case to find all the classes and their features.

Figure 10.33 shows the merged class model from the various strategies while Figure 10.34 depicts the merged class model for the CUAV Acquire Image use-case analysis. There should be no surprise from this effort, but it may be necessary to redefine an element or two to make everything consistent.

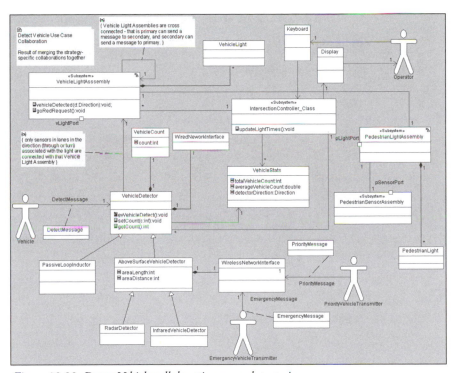

Figure 10.33 Detect Vehicle collaboration merged strategies

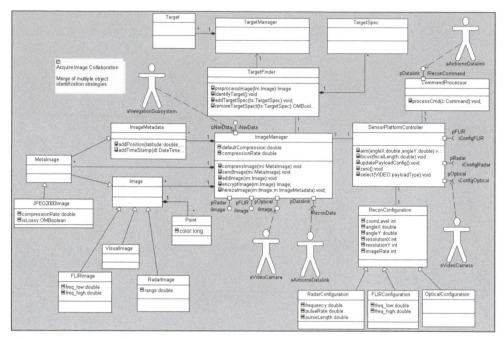

Figure 10.34 Acquire Image collaboration merged strategies

These merged class diagrams are the representation of the "analysis model" for the use case. However, in the context of this book we haven't validated that these elements are all correct. Rhapsody allows you to generate code and execute these models to ensure that they do, in fact, correctly realize the use cases. That execution, while crucial in a real project, is beyond the scope of this book. Interested readers should look over the tutorial information on the included Rhapsody installation.

Architectural Design: Answers

Answer 6.1 Concurrency and Resource Architecture

As it happens, the collaboration in Problem 6.1 includes the fixed cycle model collaboration as well. The latter collaboration is structurally simple, if behaviorally more complex. We will use the Recurrence Properties and Independent Processing as the primary task identification strategies. All the light assemblies must synchronize when one light goes from red to green or "bad things" will happen, but between these synchronization points, their processing is independent. The pedestrian lights are more tightly coupled with the traffic lights of the same orientation. Thus, as a first stab, let's make one task for traffic control for each of the two orthogonal directions of traffic. Let's use the Event Source strategy for the sensors and add interrupt handlers for the primary through traffic (north and south), primary turn traffic (north and south), and primary pedestrian traffic (two for each side of the road), and a similar set for the secondary traffic orientation. The keyboard probably runs in its own thread as well, to make sure no user input is missed, and the display probably runs in its own thread, since the display processing is independent.

Figure 11.1 is a class diagram so it doesn't show all the instances of the thread classes. Because this is a class diagram, the ports between the threads cannot be connected with links, but if it is drawn as an object diagram, those links among the ports can be added.

The UAV side of things is only slightly more complex in this exercise. This is because we are modeling threads at the subsystem level. We also haven't considered classes from other use-case collaborations. There are a couple of aspects of the collaboration that have been as of yet unconsidered. There are ports for connecting the subsystem (and its contained collaborations) with the Airborne Datalink subsystem and with the various payloads. This is likely to be done via networks, one for the UAV itself and another for connecting to the payloads. Network communications

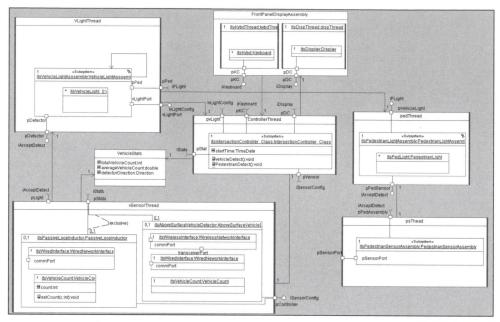

Figure 11.1 Roadrunner concurrency model

should almost always be decoupled from the execution of the required semantics, so we need two network interfaces to operate in separate threads. Additionally, the configuration of the imaging and platforms operates independently from the image processing, so the objects facilitating these functions should be in separate threads. Finally, targeting, a topic only lightly touched on so far, is clearly independent as well, and so will be in yet another thread. The concurrency model for the Reconnaissance Management subsystem is shown in Figure 11.2.

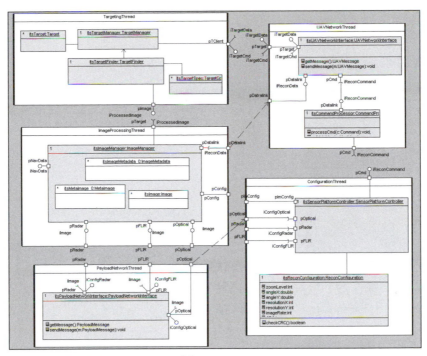

Figure 11.2 UAV concurrency model

Answer 6.2 Distribution Architecture

Figure 11.3 shows the overall distribution architecture for the Roadrunner system. The distribution-related classes are shown shaded. Note that many of the previous associations between classes have been "broken" since they will be implemented via a communication protocol. The figure shows three primary network interface classes:

1. A WiredNetworkInterface to support Ethernet connections for the front panel display (display and keyboard) and for the remote monitoring and control station.

2. A WirelessNetworkInterface to support transponder messages from high-priority and emergency vehicles and also to support wireless connections between the VehicleLightAssembly and the VehicleDetector.

3. A SerialInterface to support the wired connections among elements local to the intersection (vehicle light assemblies, vehicle detectors, pedestrian light assemblies, pedestrian detectors, and the intersection controller.

We can go a bit deeper here as well. What is the structure of the message types and other elements associated with the various interfaces? The following Figure 11.4 goes into more detail about the structure of the serial interface.

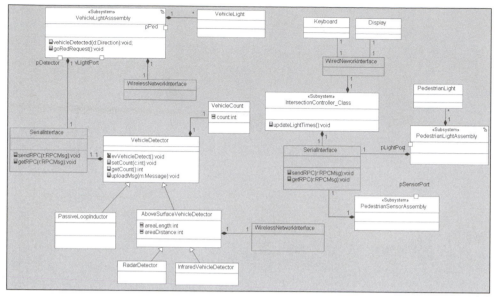

Figure 11.3 Roadrunner distribution model

Figure 11.4 shows the SerialInterface as a structured class containing a Serial-ProtocolStack, message queues (incoming and outgoing), an RPC-interface and a low-level hardware interface. The protocol stack is itself a structured class, also shown in the figure, with three layers: an application layer with interfaces to the sender

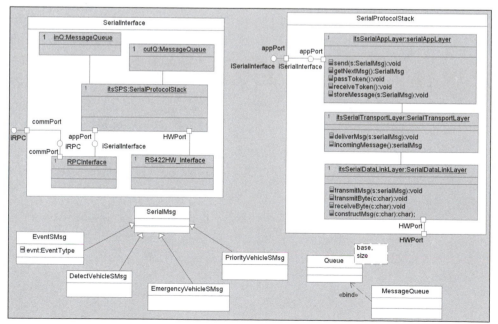

Figure 11.4 Serial interface structure

client and to clients that receive incoming messages. The structure of the messages is also shown as a generalization hierarchy in the figure.

Figure 11.5 and Figure 11.6 show the structures for the wired and wireless interfaces. They are slightly different, more to show some presentation alternatives than anything else. In a real system project, you would develop, essentially, an object analysis model for the communication infrastructure. You might also purchase a complete protocol stack that you would simply integrate into your interface classes rather than build your own.

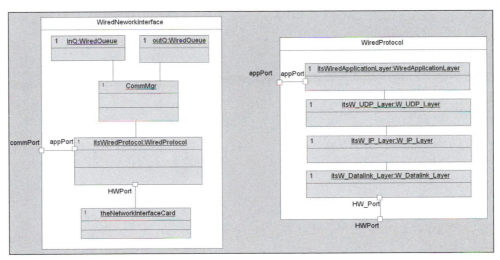

Figure 11.5 Wired interface structure

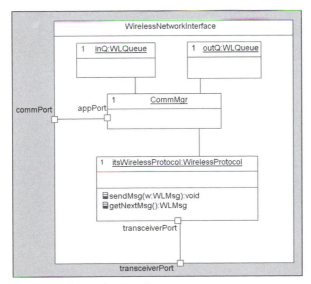

Figure 11.6 Wireless interface structure

For the UAV, we apply the Port Proxy pattern, as the problem dictated, resulting in Figure 11.7. The proxies are shown with a (slightly) different shading and with the box lines dashed. This special notation has no semantic meaning but is added simply to make the proxies more easily distinguishable. We also follow the convention of prefacing interface names with a lower-case "i".[1] Even though this diagram adheres to the "one important concept per diagram" constraint, this figure is definitely "busy" and we might very well decide to elide some of the interfaces on the diagram to make it more readable. Nevertheless, it does show how the proxies provide semantic interfaces on one side and network interfaces on the other. The proxies that connect to the payload via that Ethernet network as well as the proxies that connect to the wireless networks are also shown.

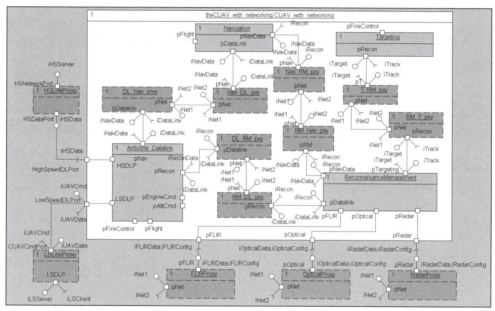

Figure 11.7 CUAV distribution with proxies

We can simplify the diagram without a loss of information by noting that if the model is well-formed, connected ports form either 1) port delegations (with compatible interfaces) or 2) port conjugates (with reversed, but compatible interfaces). That is, delegation ports connect to their targets with the same or compatible interfaces, while ports that connect between servers and clients connect with their

---

[1]   A more common naming idiom uses uppercase "I". However, in some fonts, it is difficult to distinguish uppercase "I" (upper case "i") from "l" (lower case "L") and the number "1". I prefer the lower case "i" preface for this reason.

interfaces reversed (i.e., services offered by one are required by the other, and vice versa). Since the proxies are in the middle of it all, we can remove the display of the port interfaces to simplify the diagram as shown in Figure 11.8.

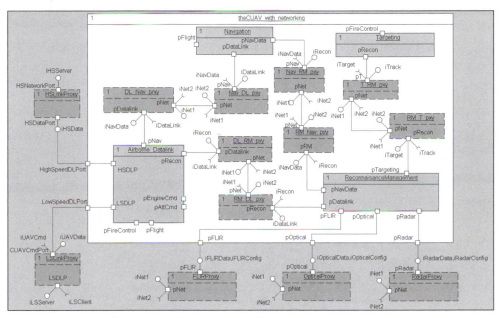

Figure 11.8 CUAV distribution with proxies (simplified)

While Figure 11.8 is still "busy," it does clearly show how the proxies glue together the subsystems over the network by providing both semantic and network-specific interfaces.

Answer 6.3   Safety and Reliability Architecture

For the Roadrunner Traffic Light Control System, the exercise involves doing the following (taken from problem chapter):

- Take the collaboration done previously (shown again as Figure 11.9) and do a separate FTA on each of the fault situations below:

    - Both primary and secondary through traffic have GREEN lights

    - Pedestrian traffic has a WALK light at the same time that the orthogonal vehicle traffic has a GREEN light

    - All lights are off

- From the result of this analysis, create a hazard analysis in which hazards and faults are identified and control measures are added, including estimated fault tolerance times

- Determine how architectural redundancy should be added to make the system single-point fault safe for the analyzed fault situations

- Draw the safety architecture class or object diagram showing how the elements collaborate to achieve the desired safety

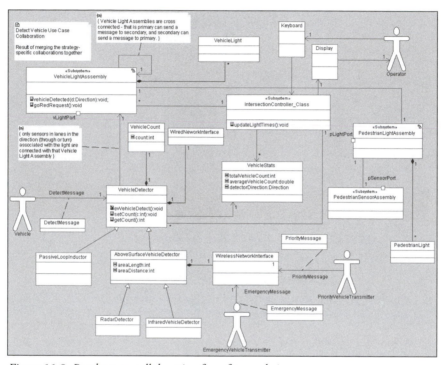

Figure 11.9 Roadrunner collaboration for safety analysis

Unfortunately, UML (and Rhapsody) doesn't have an FTA diagram editor, so I used Visio™ from Microsoft® to create the FTA diagrams. Figure 11.10 shows an FTA diagram for the "all lights green" case. The easiest way to read this diagram is from the top (the resulting hazardous condition) down; the only way to get both lights green is for one of them to be green and then the other turns green. For one of them to turn green in this case, one of three conditions must occur: either the command to the light is corrupted, the command was issued inappropriately, or a fault in the light causes it to turn green; and so on, down to more primitive fault events.

The next figure, Figure 11.11, is a fault tree for getting a green light for the road traffic in one direction while getting a walk light for the pedestrian traffic in the

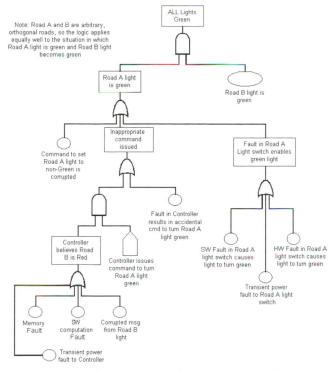

Figure 11.10 Roadrunner FTA for "all lights green" fault

orthogonal direction. The two primary conditions analyzed are 1) the pedestrian light is already WALK and the vehicle traffic light for the orthogonal road turns green (the left side of the figure), and 2) the light is already green and then the walk light turns to WALK (right side of the figure).

Figure 11.12 looks at the last analyzed condition, "All lights off." If only one light turned off, the situation might not be unsafe, if the light for the orthogonal direction turns to RED; however, the situation is clearly unsafe if all lights are off. Of course, it is not as unsafe as in the first hazardous condition analyzed (all lights green) because most people, seeing that the light is off, will slow down and proceed cautiously.[2] If they see the light is green, however, they will approach the intersection at full speed and with far less caution. We see in Figure 11.12 that either a general power failure to all lights or commands to turn off the lights (sent because of a controller or communications fault) could result in the situation in which all lights are off.

---

[2]   Except in Boston and parts of (OK, most of) Italy.

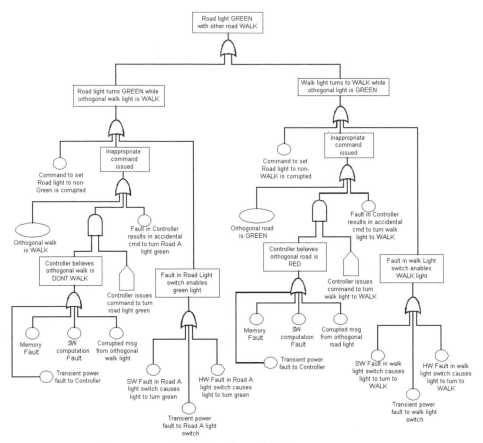

Figure 11.11 Roadrunner FTA for "walk and green" fault

The advantage of the FTA is that we can clearly see the logic of the confluence of conditions and events necessary to create the hazardous situation. It becomes simple (in principle[3]) to make the system safe by creating other conditions that must be ANDed with the identified faults and conditions to result in an unsafe condition. These "other conditions" are called "safety measures." For example, in Figure 11.12, there is a fault called "Power loss to all lights." Clearly, if the lights have no power, then the condition arises. However, if we add a UPS (uninterruptible power supply) to each individual light, that condition can't arise unless both the main power and the UPS fail. We can modify the light behavior to make the system safe by making each light go to RED while on the UPS as well as notify the remote monitoring station of the power loss; in addition, we can add "lifeticks" between the lights so that

---

[3] Remember, though, the 97th Law of Douglass, "The difference between theory and practice is greater in practice than it is in theory." And, of course, the 98th law as well: "The purpose of theory is to improve practice."

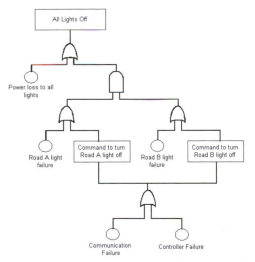

Figure 11.12 Roadrunner FTA for "all lights off" fault

if a light manifests a fault (and so doesn't respond to the lifetick), the light expecting the lifetick goes into its "fail safe state." Thus, to arrive at the unsafe condition under consideration, two independent faults must manifest themselves at the same time—this is what is meant by the term "single-point failure safe."

The hazard analysis (Figure 11.13 through Figure 11.15) below summarizes these faults and adds safety measures. These safety measures, which are typically the instantiation of safety patterns, constitute the safety architecture. This hazard analysis is shown in three tables, one per fault. In many such analyses, the hazard analysis is a single huge table of hazards, faults, and safety measures.

We see in Figure 11.13 the hazard "All Lights Green" and the various faults we identified in the previous FTA. We then elaborated the hazard analysis with control measures, and added in (what we hope are) reasonable detection and control action times. From this, the exposure time and safety are computed.

Figure 11.14 presents the same sort of hazard analysis for the hazard of having the walk lights in the WALK state while at the same time having the road light for the orthogonal road in the GREEN state. Many of the faults are the same, but others are added, because faults in the walk light system or commands can also lead to the hazard.

The last hazard analysis, shown in Figure 11.15, shows the safety analysis for the All Lights Off hazard. From this analysis, the model can be updated to include the safety measures, and the FTA can be redone. Figure 11.16 shows the FTA for the "All lights off" condition with the added safety measures. These safety measures are

Hazard Analysis for All Lights Green

Hazard	Fault	Severity (1 [low] – 10 [high])	Likelihood (0.0 – 1.0)	Computed Risk	Time units	Tolerance Time	Detection Time	Control Measure	Control Action Time	Exposure Time	Is Safe?
All lights green	Transient power fault to controller	10	0.08	0.8	seconds	2	0.01	UPS for controller; notify remote monitor	0	0.01	TRUE
	Memory fault	10	0.02	0.2	seconds	2	0.01	CRC set and checked every access	0.01	0.02	TRUE
	SW computation fault	10	0.03	0.3	seconds	2	0.1	Explicit check for unsafe conditions	0.01	0.11	TRUE
	Corrupted message	10	0.06	0.6	seconds	2	0.05	Message CRC; ACK required	0.02	0.07	TRUE
	Accidental message transmission	10	0.01	0.1	seconds	2	0.01	Buffers zeroed after memory send	0.01	0.02	TRUE
	SW fault in light	10	0.03	0.3	seconds	2	0.5	Lifeticks containing light state sent among lights; if not received, fail safe initiated and broadcast	0.01	0.6	TRUE
	HW fault in light	10	0.01	0.1	seconds	2	0.5	Lifeticks containing light state sent among lights; if not received, fail safe initiated and broadcast	0.1	0.6	TRUE
	Transient power fault in light	10	0.08	0.8	seconds	2	0.01	UPS for each light	0	0.01	TRUE

Figure 11.13 Hazard analysis for all lights green

Hazard Analysis for Walk and Green

Hazard	Fault	Severity (1 [low] – 10 [high])	Likelihood (0.0 – 1.0)	Computed Risk	Time units	Tolerance Time	Detection Time	Control Measure	Control Action Time	Exposure Time	Is Safe?
Walk light and orthogonal green	Transient power fault to controller	10	0.08	0.8	seconds	2	0.01	UPS for controller; notify remote monitor	0	0.01	TRUE
	Memory fault	10	0.02	0.2	seconds	2	0.01	CRC set and checked every access	0.01	0.02	TRUE
	SW computation fault	10	0.03	0.3	seconds	2	0.1	Explicit check for unsafe conditions	0.01	0.11	TRUE
	Corrupted message	10	0.06	0.6	seconds	2	0.05	Message CRC; ACK required	0.02	0.07	TRUE
	Accidental message transmission	10	0.01	0.1	seconds	2	0.01	Buffers zeroed after memory send	0.01	0.02	TRUE
	SW fault in light	10	0.03	0.3	seconds	2	0.5	Lifeticks containing light state sent among lights; if not received, fail safe initiated and broadcast	0.1	0.6	TRUE
	HW fault in light	10	0.01	0.1	seconds	2	0.5	Lifeticks containing light state sent among lights; if not received, fail safe initiated and broadcast	0.1	0.6	TRUE
	Transient power fault in light	10	0.08	0.8	seconds	2	0.01	UPS for each light	0	0.01	TRUE
	SW Fault in walk light	10	0.03	0.3	seconds	2	0.5	Lifeticks containing light state sent among lights; if not received, fail safe initiated and broadcast	0.1	0.6	TRUE
	HW fault in walk light	10	0.01	0.1	seconds	2	0.5	Lifeticks containing light state sent among lights; if not received, fail safe initiated and broadcast	0.1	0.6	TRUE
	Transient power fault in walk light	10	0.08	0.8	seconds	2	0.01	UPS for each light	0	0.01	TRUE

Figure 11.14 Hazard analysis for walk and green lights

Hazard Analysis for All Lights Off

Hazard	Fault	Severity (1 [low] – 10 [high])	Likelihood (0.0 – 1.0)	Computed Risk	Time units	Tolerance Time	Detection Time	Control Measure	Control Action Time	Exposure Time	Is Safe?
All lights off	Power loss to all lights	5	0.05	0.25	seconds	2	0.01	UPS to each light; lifeticks indicate liveness to other lights	0.1	0.11	TRUE
	Light failure	5	0.09	0.45	seconds	2	0.01	photodetector in light assembly detects fault; when detected broadcast fault to initiate fail safes	0.01	0.02	TRUE
	Communication failure	5	0.06	0.3	seconds	2	0.05	CRCs on messages; ACK required; state change only on ACK receipt	0.02	0.07	TRUE
	Controller failure	5	0.01	0.05	seconds	2	0.05	Explicit check for unsafe condition; Lifeticks containing light state sent among lights; if not received, fail safe initiated and broadcast	0.02	0.07	TRUE

Figure 11.15 Hazard analysis for all lights off

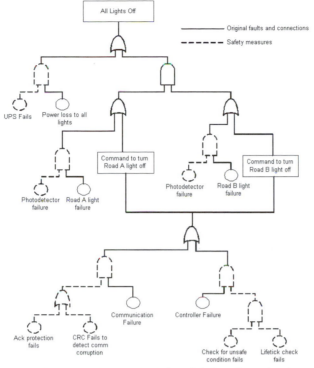

Figure 11.16 Roadrunner FTA for "all lights off" with safety measures

ultimately ANDed with the original faults events, so that to achieve the hazardous condition both the original fault must occur and the safety measure must fail.

Having completed this analysis, we can update our architecture to include the safety measures. It is interesting to note that the vehicle detectors do not have an impact on safety. If they malfunction, either by missing detections or by providing false detections, the lights will still ensure proper traffic flow. Similarly, the interaction with emergency and priority vehicles are a reliability, not a safety issue. We still want them to work and so may want to add redundancy for these elements, as we did for the lights and controller to improve system reliability, but their failure is not a safety concern. The updated Roadrunner class diagram is shown in Figure 11.17. Of course, the safety-related behavior, such as checking the CRCs and lifeticks and notifying the remote monitor of faults, isn't shown in the class diagram, so some of these behaviors are indicated with constraints.

It should be noted that we've done both a heterogeneous redundancy specialization of the Channel Architecture pattern here by providing diverse power supplies to various system elements, as well as inserting data validity checks in the manner of the Single Channel Protected pattern.

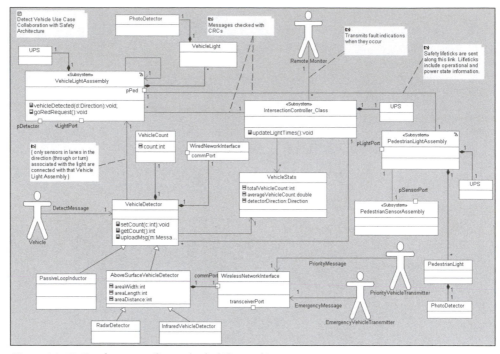

Figure 11.17 Roadrunner safety and reliability architecture

The next part of the exercise repeats this work for the Coyote UAV. Since we've focused previously on developing the reconnaissance management subsystem—involved in both target identification, tracking, and targeting—we'll continue in that vein.

- Create an FTA for the reconnaissance management subsystem that identifies the safety hazards and risks associated with this subsystem, one per hazard

- From the result of this analysis, create a hazard analysis in which hazards and faults are identified and control measures are added, including estimated fault tolerance times

- Determine how architectural redundancy should be added to make the system single-point fault safe for the analyzed fault situations

- Draw the safety architecture class or object diagram for the subsystem showing how the elements collaborate to achieve the desired safety.

For the CUAV, we'll focus on three views with respect to the safety architecture. First is the system architecture—how the overall system, including the ground station, aircraft, and payloads work together. Next is the aircraft architecture—how the aircraft subsystems interact. Lastly, we'll look inside the Reconnaissance Management

Subsystem (RMS) to see how the internals of that subsystem function, and how they relate to safety and reliability.

We will, of necessity, severely limit the scope of our analysis. In a real system, we would have to look at hundreds of hazards and even more faults, to ensure system safety and reliability. In this answer, we will limit our scope to hazards directly due to faults with the reconnaissance data or management.

To begin with, let's look at the structure of the system under consideration. Figure 11.18 shows the overall CUAV architecture including the ground system, aircraft and payloads. Figure 11.19 shows the subsystem architecture of the ground system, although not a great deal of detail about the internal structure of the control or monitoring stations. Figure 11.20 shows the subsystem architecture of the aircraft itself. Lastly, Figure 11.21 shows the internal structure of the RMS.

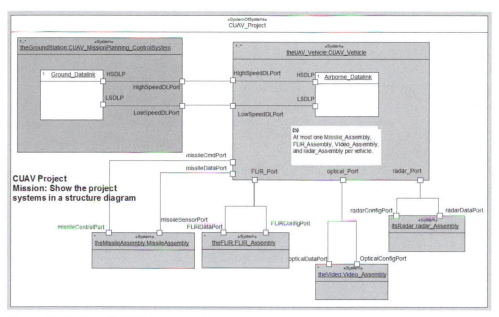

Figure 11.18 Overall CUAV system architecture

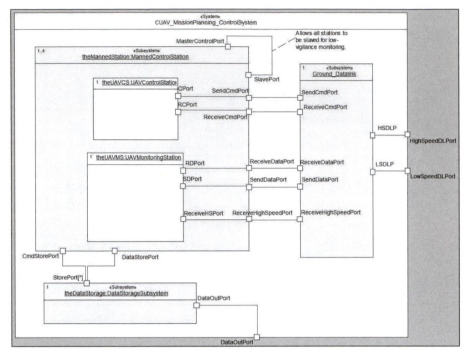

Figure 11.19 CUAV ground system architecture

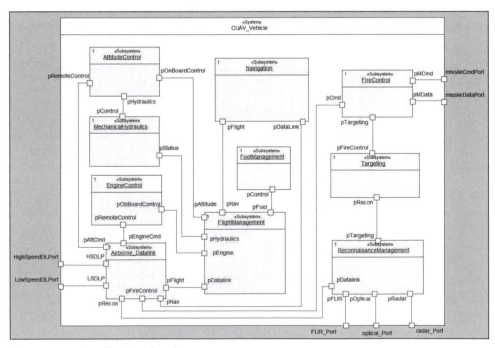

Figure 11.20 CUAV aircraft architecture

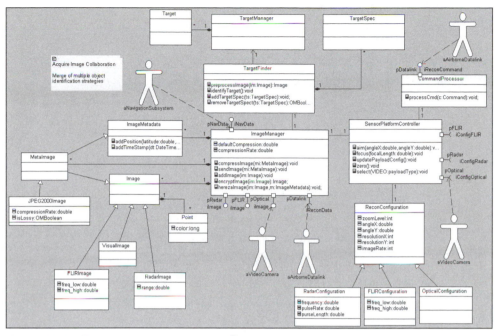

Figure 11.21 Reconnaissance management subsystem structure

Given that structural overview, let's now list the hazards we are specifically addressing:

- target misidentification (i.e., "false positive")
- lack of (valid) target identification (i.e., "false negative")
- loss or corruption of target information either in storage or transmission, e.g.,
  - incorrect target classification or Combat ID (CID)
  - incorrect location or coordinates
  - loss of information about target
  - incorrect target kinematics (vector, flight path, etc.)
  - incorrect target management (e.g., failure in sensor fusion)

We have focused previously on the reconnaissance management and not the targeting system, so it makes sense to understand how the responsibilities of the RMS and targeting subsystems differ. The RMS does low-level surveillance of the physical environment and then identifies elements in the environment that are likely to be targets of interest. This is the job of the TargetFinder class. The TargetFinder has a set of TargetSpecs, each of which identifies the observable characteristics of something to look for—size, shape, image template, color, speed, etc. Once a target is identified and classified by the TargetFinder, it is put into the list of targets, managed by the TargetManager. When appropriate, it can be passed to the Targeting subsystem as a target at which to fire a missile. The FireControl subsystem is responsible for guiding the missile to the target under Mission Ops personnel direction.

These faults can be analyzed with FTA. Figure 11.22 looks at the Target Misidentification and Failure to Identify Target hazards; to save space, the safety measures have already been added (as dashed lines) to this FTA. The primary culprits that could lead to misidentification, either false or negative, are:

- Communication corruption
- Memory corruption
- Computation fault
- Noise-reduction algorithm is insufficient to resolve the noise

Other than the last, these issues are addressed by adding CRC protection to messages to be transmitted and to data in situ, on a per-class basis. This protection must be added to communication links within the aircraft, within the ground station, and to the command (low-speed data) and reconnaissance (high-speed data) links between the two.

Note the use of the single-channel protected pattern in the inclusion of lightweight data validity checks. Heterogeneous redundancy improves safety through the use of two computational paths (using different algorithms) for the computation of target classification, location, kinematics, and CID. In addition, the problem itself also offers heterogeneous redundancy in that it has two wireless links. It is expected that the high-speed link will fail due to jamming or noise before the low-speed link. The low-speed link is to be designed with broad spectrum and built-in redundancy (e.g., CRCs) and encryption to handle security. The high-speed link does not have these measures because 1) there isn't enough bandwidth to add much redundancy, but 32-bit CRCs are added to the video frames for error detection, 2) the loss of data isn't as critical as undetected corruption, and 3) practice has shown that real-time telemetry interception is not a security risk, since the enemy presumably already knows where they are ;-).

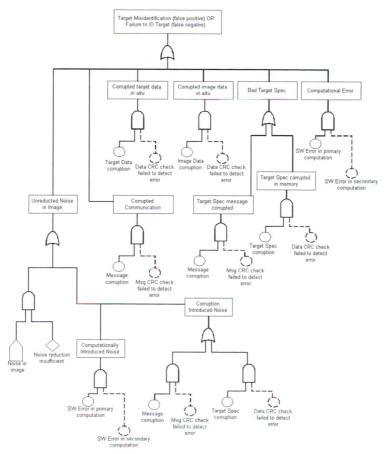

Figure 11.22 CUAV FTA for target misidentification

Figure 11.23 shows how the heterogeneous redundancy and single-channel protected patterns are applied to the RMS:

- Heteregeneous redundancy pattern
  - TargetFinder is replicated using different algorithms to perform preprocessing and target identification (which also computes CID)
- Single-channel protected pattern
  - Command processor validates messages with CRCs
  - ImageManager uses CRCs on navigation data messages from the Navigation subsystem (not shown on diagram)
  - Targets are stored with one's complement (bit-inverted) copies of location and direction attributes

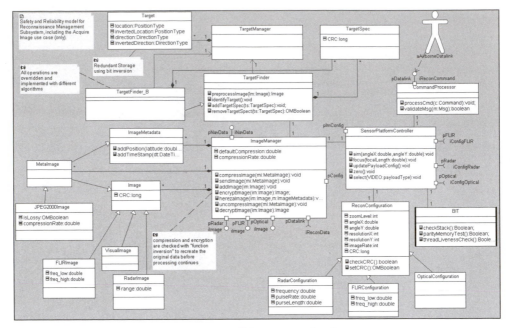

Figure 11.23 Reconnaissance management collaboration with safety patterns

- Image superclass contains a CRC and (not shown) set and check CRC operations which are overridden in the subclasses

- ImageManager uses algorithmic inversion to "back compute" the original data from the computation for both compression and encryption operations

- ReconConfiguration uses a CRC and set and check operations to validate data

- BIT class performs built-in-test procedures in a separate thread to check the stack (using "high-water marks"[4]), a parity memory test (supported by parity in hardware), and checks that all threads are running and issuing lifeticks at a minimal rate.

The architectural diagrams have not been changed to account for the safety architecture because the error detection takes place primarily within the subsystems (such as the RMS) and at the message level between subsystems, using protection means on each message. If there is data loss because a subsystem goes down, then the mission ops actor will take appropriate action, such as reverting to another air vehicle.

---

[4]   The stack "high-water mark" test simply fills the stack for each task with a known pattern (such as "A5") and checks periodically that these values have not been overwritten above (or below, as appropriate) a certain position in the stack.

Mechanistic and Detailed Design: Answers

Answer 7.1   Applying Mechanistic Design Patterns—Part 1

For the Roadrunner portion of this exercise, the application of the adaptor pattern is very straightforward. The design class diagram is shown in Figure 12.1.[1] The Vehicle-Detector class is changed from a class to an interface. The adapters subclass both from a specific implementation class of the specific sensor type and the iVehicleDetector

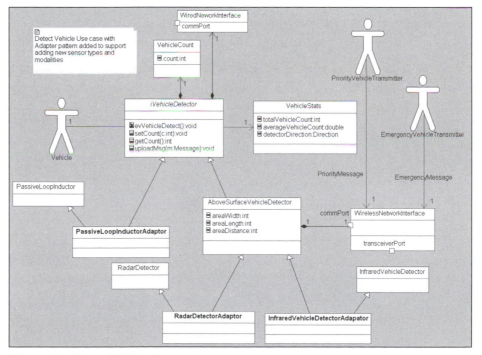

Figure 12.1 Detect Vehicle adaptor pattern

---

[1]   Parts of the collaboration that weren't affected are elided from the diagram for simplification.

interface. The operations from the iVehicleDetector interface are then elaborated into methods that invoke the specific services provided by the various sensors.

The CUAV problem required the application of the chain of responsibility pattern to make the use and addition of new image-processing algorithms easier and more flexible. The first question is where the image processing takes place. Two obvious candidates are under the control of the ImageManager or under the control of the TargetFinder. The ImageManager class is responsible for early image processing, which would naturally include sharpening and enhancement of the images. On the other hand, the TargetFinder might employ different enhancement methods to find different kinds of targets, so an argument can be made for both. It is even possible to apply the CoR pattern twice, once for basic image enhancement and again for target-specific enhancements.

In this case, we'll apply the pattern to the ImageManager only. In terms of image processing, we'll include not only image enhancement, but also conversion to a compressed (JPEG) format, encryption, averaging, and center-surround sharpening as image-processing steps. We also have a need to undo these steps, so we've applied the pattern twice, once to do image processing and once to undo some of the steps, such as encryption and compression. As with the Roadrunner pattern application, we only show the relevant portion of the collaboration.

You'll note in Figure 12.2 that some of the operations defined for the ImageHandler, such as compression and encryption, have been removed since they are now handled

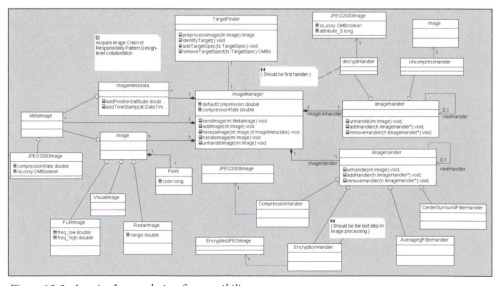

Figure 12.2 Acquire Image chain of responsibility pattern

by the image handlers in the chain of responsibility. This small reorganization of responsibilities is an example of the refactoring we referred to previously.

Answer 7.2 Applying Mechanistic Design Patterns—Part 2

The observer pattern optimizes the interaction between client and server so that additional clients can be added, and so that data is only sent when it is appropriate to do so. Each light assembly coordinates the traffic for a single road and direction for both through and turn traffic, so there are two light assemblies per road, for a system total of four. There are also a total of eight pedestrian light assemblies (two per road per side). All these light assemblies must share information to ensure that the overall system is in a safe state at all times (specifically cross-oriented traffic does not have non-red or non-Don't Walk at the same time). Figure 12.3 is an object diagram showing the connections among the light assembly instances. The instance naming convention assumes the primary road is north-south and the secondary road is east-west. Thus, the name for the vehicle light instance on the primary road that faces north is "vPrimaryNorthFacingLight" and the walk light for the secondary road

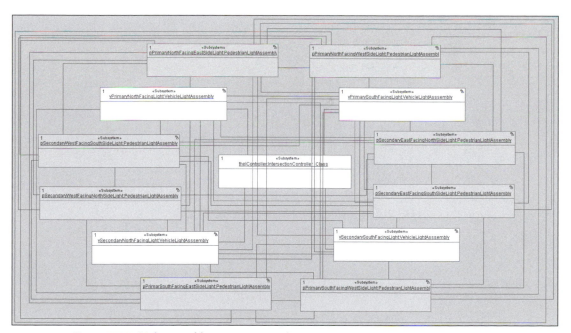

Figure 12.3 Light assembly connection topology

---

2   Extra credit to the reader who validates all the links. ☺

that faces east and is on the north side of the street is name "pSecondaryEastFacing-NorthSideLight. It's verbose, but has the advantage of being unambiguous.

Figure 12.3 shows the link topology of the system.[2] The intersection controller is in the middle, the white-filled objects are the vehicle lights, and the shaded boxes are the pedestrian lights. As you can see, the topology is complex and validating that all the lights are connected to all the correct clients would be a chore.

When we apply the data bus pattern to all the lights and clients (they are largely the same, after all), the topology is much more straightforward, because the lights and clients need only a single link to the data bus instance. The responsibility for the topology creation is distributed to the clients and servers. Each light assembly is a server, so it must link to the data bus instance and call update() to update its light state. Each client must subscribe to all the other lights. Note that the intersection controller is the special case because it is not a light but is a client to all the light assemblies.

The data structure is simple, nothing more than an enumerated value containing all of the possible states for the light. Two different enumerated types are needed, one for vehicle lights and one for pedestrian lights.

Each light assembly then maintains a (local) list of all the relevant (cross-traffic) lights. When those lights change state, the data bus updates the light assembly, which in turn updates its (local) copy. The behavior of each light then checks consistency with the other lights, something like this:

```
//wants to change state to permit traffic to flow (i.e. yellow or green)
Set a global state change semaphore (also in the data bus, but updated by
all lights)
For all cross traffic vehicle lights
    If it is not red,
        send an error message to the intersection controller
        release the global semaphore
        exit
For all cross traffic pedestrian lights
    If it is not Don't Walk
        send an error message to the intersection controller
        release the global semaphore
        exit
Change state
Release the global semaphore
```

This can be optimized somewhat by observing that certain light sets change together; for example, the walk signals that face each other on the same road and side of the street change together. Such light sets can be controlled by a single state change. However, the light set construction may be nontrivial. For example, it is not necessarily true that through lights facing different directions on the same road change together; if the turn lanes are operated in SEQ mode, then the turn light in one direction is set when the through lanes in the same direction are on but the opposing traffic on the same road must be prevented from going at that time. So, in some cases, such light sets are mode-specific.

Figure 12.4 shows the Detect Vehicle collaboration updated to include the data bus pattern. Note that the shared data are the light states. There are two data classes represented, one for the vehicle light states and one for pedestrian light states. Note that the clients and server light assemblies are replicated on the diagram—this is simply to minimize line crossing. On the server side, they have a composition with a unary multiplicity to the data they are providing, while on the client side, they have a * multiplicity to the data they get from their peers.

The dataID is an enumerated type that identifies the road (PRIMARY or SEC for secondary), whether it is THROUGH or TURN (vehicle lights only), whether it faces north, south, east, or west (N, S, E, W) and, in the case of the pedestrian lights, which side of the road it's on (the last character N, S, E, or W). This differentiates the data from one server from all the others. Note that there will be at least one instance of light state data for each enumerated value in the IDType.

The global semaphore was added as an attribute of the data bus because it affects all data—a change in the state of a single light is fundamentally a change in the system

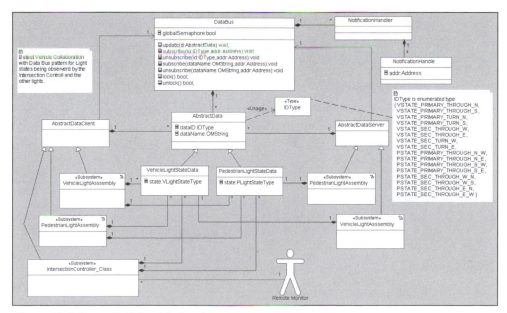

Figure 12.4 Detect Vehicle with data bus pattern

state. The lock() operation returns TRUE if the attempt to lock was successful and FALSE otherwise; similarly, the unlock() operation returns TRUE if the attempt to unlock was successful and FALSE otherwise (although this should never happen unless the system contains an error).

While the topology shown in Figure 12.4 may seem complex, in reality it is an order of magnitude simpler than the topology shown in Figure 12.3 at the cost of some additional objects and the memory that they require.

The second half of this problem requires the command pattern be applied to the Acquire Image collaboration. The command pattern is straightforward and requires that the command hierarchy be identified, as well as how the commands invoke services on the existing elements in the Acquire Image collaboration. In addition, it is necessary to specify how the command objects come into existence.

The solution is shown in two different class diagrams. The first, Figure 12.5, shows the generalization taxonomy for the command classes and how the commands are created. Specifically, they get passed to the command processor via the TCP/IP Protocol Stack encapsulated into datagrams, one command per datagram. Once the datagram is received, the command is extracted and placed into a command queue for processing. The commands themselves are specialized into two kinds—configuration commands and direct requests for actions or data.

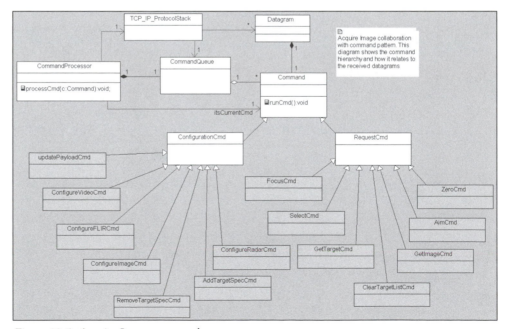

Figure 12.5 Acquire Image command structure

Each of the specific commands is a specialized form of command and contains relevant data and a specialization of the execute() command to invoke the appropriate service of the elements of the collaboration. The command processor has the responsibility to create the link between the command instance and the appropriate collaboration object and then invoke the command's runCmd() method. Figure 12.6 shows the associations from the various commands (shaded for emphasis) to the collaboration classes.

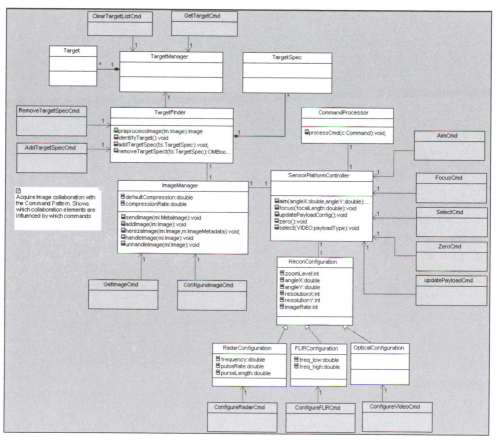

Figure 12.6 Acquire Image command pattern

Answer 7.3 Applying Detailed-Design State Behavior Patterns

The any state pattern is usually simple to apply and the situation with the light assembly state machine is no exception. Before we look at the solution, let's look at what the high-level state machine would look like if we don't use the pattern. Figure 12.7 shows the resulting rat's nest. Note that, except for the Off state, each of the states in the figure is a composite state containing a possibly complex state machine inside. Because the depiction of those state machines is deferred to a separate diagram, they are referred to as "submachines."

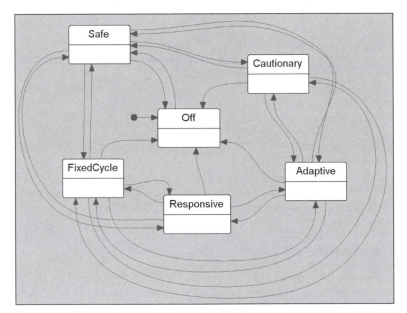

Figure 12.7 Light assembly high-level statechart before any state pattern

With the applied pattern, the state machine is simpler and far easier to understand, as you can see in Figure 12.8. The light assembly can transition from any mode to any other mode directly, just as in the previous figure, but this one can actually be read!

As mentioned, each of the states in Figure 12.8 holds a separate state machine for specifying the particular behavior of that operational mode. We've only detailed one of these so far, the fixed cycle mode. Rhapsody provides a simple shortcut for navigating to that submachine; simply click the "submachine icon" contained within the state and the submachine diagram opens. The submachine for fixed cycle mode is shown in Figure 12.9.

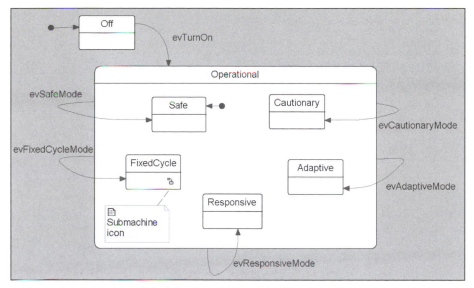

Figure 12.8 Light assembly high-level statechart with any state pattern

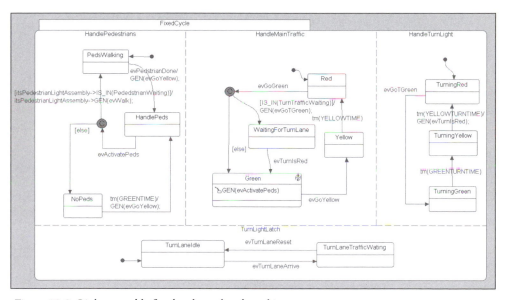

Figure 12.9 Light assembly fixed cycle mode submachine

For the CUAV part of this problem, the first step is to make the ImageManager a composite class containing the device drivers as parts. Figure 12.10 does this and links the appropriate device driver to the correct ports.

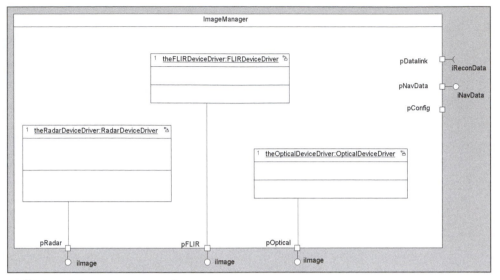

Figure 12.10 ImageManager with parts

Now, we want to specify how the classes of these parts relate to each other; specifically, how they are subclasses of a base class called SensorPayloadDeviceDriver. Figure 12.11 shows this. Note, by the way, the small state icon in the upper righthand corner of the classes. The model includes a state machine for the SensorPayloadDeviceDriver class and because the specific device-driver classes are subclasses of this base class, they automatically inherit its state behavior.

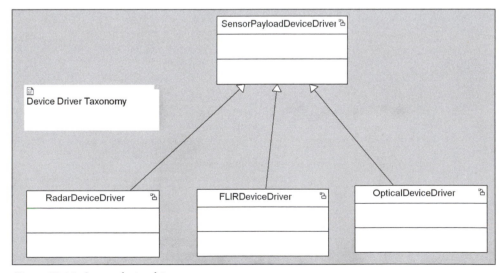

Figure 12.11 Sensor device driver taxonomy

The next step was to elaborate the state machine for the SensorPayloadDevice-Driver. This state machine is shown in Figure 12.12. Note the use of the any state pattern to handle the desired semantics of the evDisableSensor and evImageStart events. The state machine uses both actions on transitions as well as state entry actions. If desired, the state machine can be written using only actions on transitions or only state exit actions.[3]

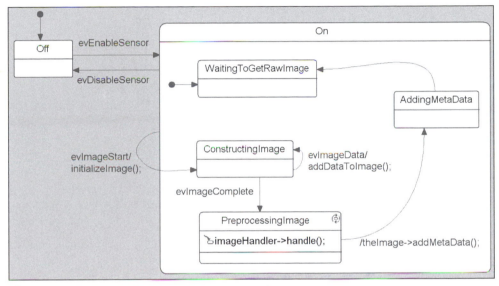

Figure 12.12 Sensor payload device driver state machine

The last aspect of this problem is to add the polling state pattern to this state machine to separate the semantic actions and states (shown in Figure 12.12) from the polling behavior. The result is shown in Figure 12.13.

This is a relatively straightforward application of the polling state pattern. The upper and-state replicates the semantics of the previous state machine. The lower and-state realizes the polling behavior. The use of the conditional pseudostate allows different transition branches to be taken based on the result of the condition of the success value returned from the acquireData() operation. Note the use of an anonymous state (state_8). Why is that there?

UML state machines provide what is called run-to-completion behavior. What this means is that each transition, including the guard, predecessor state exit actions, transition actions, and subsequent state entry actions are run completely before the

---

[3] Exercise for the reader … ☺

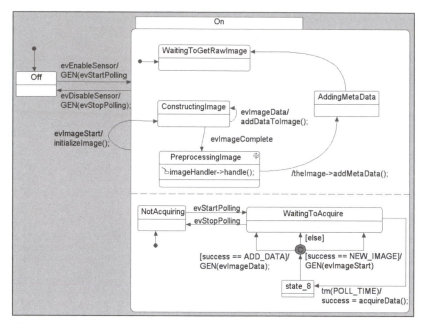

Figure 12.13 Sensor payload device driver with polling state pattern

next event is processed. However, the guards are executed first before any actions are executed. This means that if the conditional pseudostate is connected to the tm() transition, the value of success will be used before it is set by the execution of the acquireData() action. We need some way to force the execution of this action before the guards are evaluated. One solution, used in Figure 12.13, is to put a state in between the conditional pseudostate and the tm() transition. Because the tm() transition terminates on a state, all the actions are executed. The transition from state_8 to the conditional pseudostate is null-triggered, so it fires immediately after the run-to-completion step. Since the previous step assigned success a value, then it can be used in the guards.

Note, however, a possible pitfall. If success returns a value other than those listed in the exiting transitions then the state machine would be stuck in state_8 because there would be no correct exiting transition. For this reason, we've included an *else* transition guard to handle both the case NONE and any unexpected returned value.

Additional Work

1. Fill out the remaining states for different light modes of Safe, Cautionary, Adaptive, and Responsive by creating new submachines for those nested states.

2. Find other classes in the collaboration that are "reactive" (i.e., exhibit state behavior) and construct state machines for them.

Answer 7.4 Applying Detailed-Design Idioms

The solution to this problem lies in creating state machines for the lights that can generate or handle the appropriate exception events and go to the appropriate failsafe state when a fault occurs. The first state machine is for the vehicle light, shown in Figure 12.14. The isOnGreen(), isOnYellow(), and isOnRed() operations invoked the services of the appropriate photo detector and return TRUE if it detects that the specified light is on and FALSE otherwise. The exNoGreen, exNoYellow, and exNoRed events represent the exceptions that are raised if a particular light is out. The appropriate action in this case is to go to the Red state; if that is the light that is out, then no safe action can be taken.

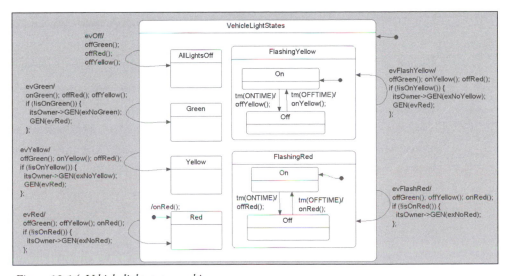

Figure 12.14 Vehicle light state machine

As can be seen in Figure 12.15, the pedestrian light state machine is very similar, if slightly simpler.

The next two figures, for the vehicle light assembly and pedestrian light assembly, respectively, handle the events as appropriate. However, rather than sprinkling the behavior throughout the detail submachines of the state machines, it is appropriate to handle it at the highest level. The first of these is the Vehicle Light Assembly state machine, in Figure 12.16.

A fault state has been added; it is differentiated from the Safe state so that the data bus can easily indicate to the clients the situation. The data bus pattern IDType must be elaborated to include the new fault state so that clients can detect that it

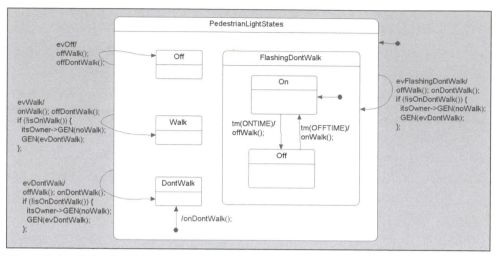

Figure 12.15 Pedestrian light state machine

went into the fault state. Also notice that while the vehicle light identifies a number of different faults, only one shows up here. Events are generalizable in the UML.[4] The exLightFault is defined in the model as the base class for all the light faults, both vehicular and pedestrian. This means that only a single transition must be represented in the client state machine, as long as the treatment of each is the same (which in this case, it is).

The astute reader might notice that it isn't necessary to use the and-state in the vehicle light assembly state machine. We could have just used the exLightFault event directly on the main semantic state machine. While this is true, we expect more fault detection and handling as the detailed design progresses and applying the pattern here will have a beneficial effect later, as more fault events and conditions are identified. If this isn't the case, the state machine can easily be simplified downstream by removing the and-state and changing the transition going to the fault state to be triggered by the exLightFault event.

The last state machine, shown in Figure 12.17, is for the pedestrian light assembly. This state machine incorporates the exception state pattern in the same way as the vehicle light assembly.

---

[4]   There is no way to draw this in Rhapsody, but you can specify it. Use the generalization relation among events in the Features Dialog for the event.

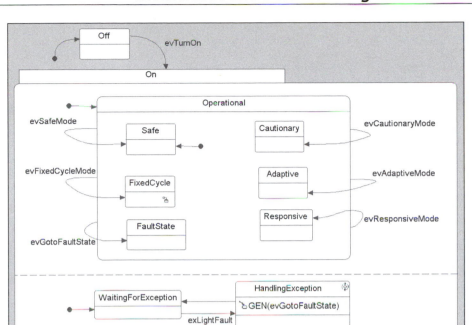

Figure 12.16 Vehicle light assembly state machine with exceptions

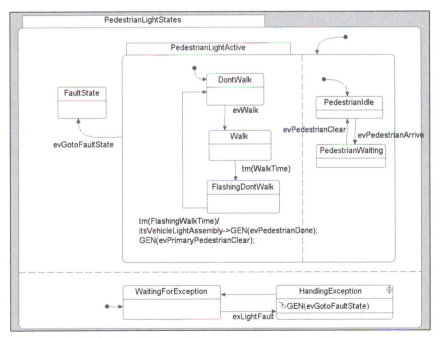

Figure 12.17 Pedestrian light assembly state machine with exceptions

The CUAV problem is arguably mechanistic rather than detailed design, but is nevertheless appropriate for the chapter and the application in question. The application of the pattern is very straightforward. Figure 12.18 shows the various sensor device drivers connecting to ports. These ports will be connected to an instance of the Watchdog class at run-time. This is implied by the association from the ImageManager Class near the port. Because the device drivers must be a subclass of WatchdogClient, this is shown in a separate diagram, Figure 12.19. The safety executive has an association to the ImageClient and so may destroy it and create a new one should the watchdog fire. If desired, the safety executive could also signal the Reconnaissance Management subsystem to perform that task as well.

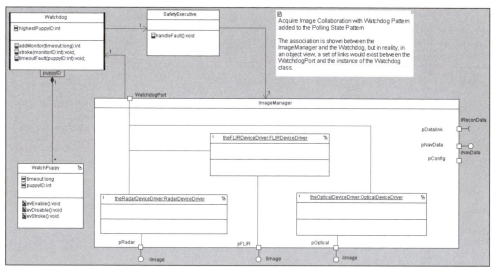

Figure 12.18 Acquire Image collaboration with Watchdog pattern

The state machine for the WatchPuppy is unchanged from the pattern specification, but is shown in Figure 12.20.

Lastly, the SensorPayloadDeviceDriver's state machine is modified to stroke the Watchdog. Because the device driver polls periodically, that is a convenient place to add the Watchdog stroke behavior. The updated state machine is shown in Figure 12.21.

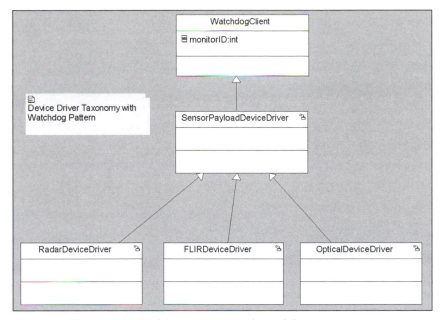

Figure 12.19 Sensor device-driver taxonomy with watchdog pattern

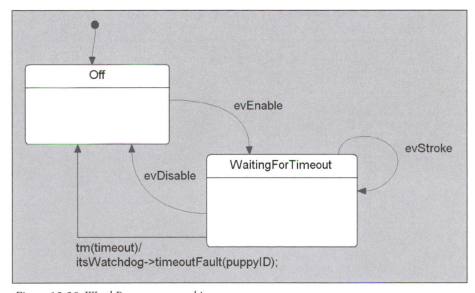

Figure 12.20 WatchPuppy state machine

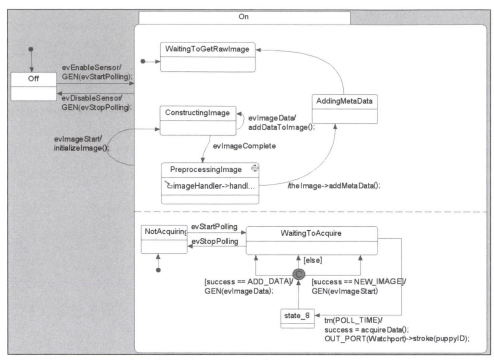

Figure 12.21 Sensor payload device-driver state machine

The Roadrunner Intersection Controller System Specification

Overview

The Roadrunner™ Intersection Controller (RIC) is an automobile intersection controller that controls individual traffic lights, subsurface and above-surface vehicle detectors and pedestrian button signal sources for a single intersection. Each intersection is limited to two intersecting roads, and supports both pedestrian and vehicular traffic in both directions along the road; options include one-way roads and turn lanes. The RIC can be programmed from a front panel display (FPD) or a remote traffic manager via a wired Ethernet interface, such as the Coyote Traffic Management System, available separately.

The Intersection Controller (IC)

Each intersection is controlled by a separate intersection controller (IC) that may be tuned manually from a secured front panel or through a remote network connection. The intersection controller supports up to six vehicular lanes (three in each direction, including a turn lane), which may be subsurface passive induction loop, or above surface infrared or radar detectors. Pedestrian lights and buttons are supported in each direction. Initial setup of the intersection controller shall configure the number of sensor sources for the intersection.

Each intersection controller shall have a panel control that allows direct local configuration and mode setting.

In addition to all normal operational modes, the RIC shall have a parameter to respond to both priority and emergency vehicle transponders. Priority vehicle transponders are used primarily for mass transit vehicles (e.g., busses) and allow the optimization of their schedules. Emergency vehicle transponders indicate an approaching emergency vehicle (typically fire and police agencies). Such transmitters shall be highly directional so that it is possible to identify which road (primary or secondary)

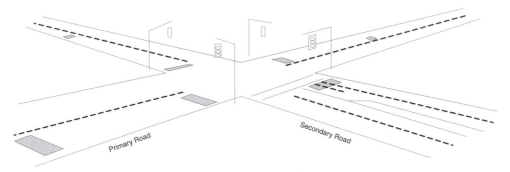

Figure A.1 Intersection

and which lane the vehicle is in. Both priority and emergency modes are only available with the above surface infrared and radar vehicle sensor configurations.

For priority vehicle transmitters, when activated and when the Priority Active parameter is set to TRUE, then when the transponder is within range of the intersection, the intersection shall either extend the through traffic GREEN by 10 seconds or, if the through traffic light is RED, shall shorten the cross-traffic light green by 10 seconds. This is to expedite priority traffic through the intersection.

For emergency transponders, when activated and when the intersection has its Enable Emergency Traffic parameter set to TRUE, then when the transponder is within range of the intersection, the intersection shall immediately cycle the lights to RED for cross traffic and GREEN to same-direction traffic; with all turn lanes set to RED. If the same-direction traffic is already GREEN, then the GREEN time shall be extended. The light shall stay in the emergency state (GREEN for same direction, RED for cross direction) until 5 seconds after the transponder has passed the intersection or been disabled. Any intersection traffic priority processing shall be aborted in the presence of an active emergency transmitter.

Configuration Parameters

A number of configuration parameters may be set that apply to all or many modes. Parameters specific to a particular mode are described within that mode. All parameters may be set on the front panel or set via the RIC, if present.

Table A.1

Parameter	Value range	Description
Commanded Mode (CMp)	0..5	Sets the currently active mode (default 0)
Primary Road (PRp)	0 .. 1	Identifies which road is primary (default, first inputs)
Primary road directions (PRD)	SINGLE, DUAL	Identifies if the primary road is one-way (SINGLE) or two-way (DUAL) (default DUAL)
Secondary road directions (SRD)	SINGLE DUAL	Identifies if the secondary road is one-way (SINGLE) or two-way (DUAL) (default DUAL)
Vehicle detector type (VDTp)	NONE, SPLI, ASI, ASR	Identifies which kind of vehicle detectors are used. Note: all active lanes within an intersection must use the same type of vehicle detector (default NONE)
Wireless Frequency (WFp)	0..10	Selectable from 10 wireless frequencies (default 0 – NONE)
Primary Turn Lanes (PTLp)	FALSE, TRUE	Indicates whether primary road has separate turn lane detectors (default FALSE)
Secondary Turn Lanes (STLp)	FALSE, TRUE	Indicates whether secondary road has separate turn lane detectors (default FALSE)
Turn Lane Mode (TMp)	SIM, SEQ	If the turn mode is SIM, then turn lane lights for both directions activate simultaneously; if the mode is SEQ, then the turn lights for both directions are done sequentially, and the turn light GREEN occurs at the same time as the through light GREEN.
Primary Pedestrian (PPp)	FALSE, TRUE	Indicates whether primary road has pedestrian buttons and walk indicators (default TRUE)
Secondary Pedestrian (SPp)	FALSE, TRUE	Indicates whether secondary road has pedestrian buttons and walk indicators (default TRUE)
Priority Active (PAp)	FALSE, TRUE	When TRUE, RIC receives and responds to the presence of signals from priority transponders (note: only valid with above-surface infrared and radar vehicle detectors) (default FALSE)
Emergency Active (Eap)	FALSE, TRUE	When TRUE, RIC receives and responds to the presence of signals from emergency priority transponders (note: only valid with above-surface infrared and radar vehicle detectors). (default FALSE)
Current Time	Hh:mm:ss	Current time of day in 24-hr format (no default)
Current Date	M:D:Y	Current date month:date:year (no default)
Morning Start	Hh:mm	Start of morning rush mode (default 06:00)

Table A.1 (continued)

Parameter	Value range	Description
Morning Mode	Mode	Mode of the intersection for morning rush (default 0)
Midday Start	Hh:mm	Start of midday traffic mode (default 10:00)
Midday Mode	Mode	Mode of the intersection for midday traffic (default 0)
Evening Start	Hh:mm	Start of evening rush mode (default 16:00)
Evening Mode	Mode	Mode of the intersection for evening traffic (default 0)
Night Start	Hh:mm	Start of night mode (default 21:00)
Night Mode	Mode	Mode of the intersection for night traffic (default 0)

The intersection shall be able to perform vehicle counting and produce the following performance statistics:

Table A.2

Parameter	Description
Primary vehicle count	Number of vehicles that have passed through the intersection since manual reset (primary road)
Primary traffic count morning	Count of vehicles through the previous morning period, or the current morning period if active
Primary traffic count midday	Count of vehicles through the previous midday period, or the current morning period if active
Primary traffic count evening	Count of vehicles through the previous evening period, or the current morning period if active
Primary traffic count night	Count of vehicles through the previous night period, or the current morning period if active
Secondary vehicle count	Number of vehicles that have passed through the intersection since manual reset (secondary road)
Secondary traffic count morning	Count of vehicles through the previous morning period, or the current morning period if active
Secondary traffic count midday	Count of vehicles through the previous midday period, or the current morning period if active
Secondary traffic count evening	Count of vehicles through the previous evening period, or the current morning period if active
Secondary traffic count night	Count of vehicles through the previous night period, or the current morning period if active

Intersection Modes

The RIC supports a number of different operational modes. Modes may be set on the secured front panel for the intersection, or they may be set by the RIC.

Mode 0: Safe Mode

This mode is the default when the system has not been configured. All vehicle lights outputs are set to FLASHING RED and pedestrian lights are disabled. This mode persists until another mode is selected. Flash cycle time is 75% "on" duty cycle on at a rate of 0.5 Hz.

Mode 1: Evening Low Volume Mode

In this mode, the designated primary road is set to FLASHING YELLOW, secondary road is set to FLASHING RED, and pedestrian lights are set to off. Same cycle times as in Mode 0 shall be used.

Mode 2: Fixed Cycle Time

Mode 2 is the most common operational mode. In this mode, the lanes cycle GREEN-YELLOW-RED in opposite sequences with fixed intervals. The system shall ensure that if any traffic light is non-RED, then all the lights for cross traffic shall be RED and pedestrian lights (if any) shall be set to DON'T WALK. Note that the turn lane times and/or pedestrian times are only valid in this mode if (1) the turn lane and/or pedestrian parameter is set TRUE in the RIC system parameters and (2) if a signal from the appropriate detector determines the existence of waiting traffic for the turn or pedestrian light.

The durations of the light times shall be independently adjustable by setting the appropriate parameters (see below). Note that, in the table, the values in parentheses are defaults.

Table A.3 Mode 2 parameters

Parameter	Value type	Description
Reset Parameters	FALSE, TRUE	(FALSE) Sets all the parameters for Mode 2 to defaults
Primary Green Time (PG2)	10 to 180 seconds	(30) Length of time the primary green light is on
Primary Yellow Time (PY2)	2 to 10 seconds	(5) Length of time the primary yellow light is on
Primary Red Delay Time (PR2)	0 to 5 seconds	(0) Length of time between when primary red light is turned on and the secondary green light is activated

Table A.3 Mode 2 parameters (continued)

Parameter	Value type	Description
Primary Walk Time (PW2)	0 to 60 seconds	(20) Length of time the primary WALK light is on when the primary GREEN light is activated
Primary Warn Time (PA2)	0 to 30 seconds	(10) Length of time the primary FLASHING DON'T WALK light is on after the WALK light has been on
Primary Turn Green Time (PT2)	0 to 90 seconds	(20) Length of time the primary turn light is GREEN. Note: only valid when the Primary Turn Light parameter is TRUE.
Primary Turn Yellow Time (PZ2)	0 to 10 seconds	(5) Length of time the primary turn light is YELLOW. Note: only valid when the Primary Turn Light parameter is TRUE.

The default values depend on the system configuration.

Table A.4 Default cycle times for Mode 2

Turn Lane	Ped Signal	Green	Yellow	Red	Walk	Don't Walk	Turn Green	Turn Yellow
F	F	30	5	0	0	0	0	0
T	F	30	5	0	0	0	15	5
F	T	50	5	0	15	5	0	0
T	T	50	5	0	15	5	15	5

The values in Table A.4 are true for each direction, independently. Thus, if the primary road has a car waiting in its turn lane and a pedestrian walking, but the secondary road has neither, then the following timing diagram represents the cycle times for simultaneous turn lane mode (i.e., the turn lanes in both directions for a road turn together and the straight traffic doesn't begin until the turn lanes have cycled to Red).

Scenario: Pedestrian and turn lanes enabled 001

Preconditions: Secondary light is GREEN, Primary is RED; Car waiting in primary turn lane; pedestrian waiting for primary walk signal; 5s left in Secondary cycle GREEN

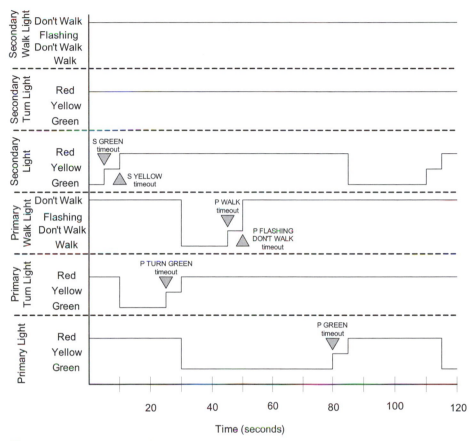

Figure A.2 Roadrunner Mode 2 timing diagram

Mode 3: Responsive Cycle Mode

Mode 3 provides a fixed cycle time when secondary road is triggered by either pedestrian or vehicle. That is, it operates exactly like Mode 2 except that in the absence of cross-traffic signals (vehicle or pedestrian), primary road or primary pedestrian signals, the system shall cycle to the default condition: primary through lights are GREEN, primary turn lights are RED, secondary through and turn lights are RED, and all pedestrian lights DON'T WALK. As long as the signal is refreshed within the GREEN or WALK times, the currently active vehicle or walk light shall be refreshed. If the appropriate interval elapses without a fresh vehicle or pedestrian signal, then the intersection shall cycle to the default state. The default shall be maintained for at least the minimum duration of the Primary Green Time , as specified by the Mode 3 parameters. If there are both turning vehicles and pedestrians waiting along the same road, they shall be handled in same order as in Mode 2 (turn lane first, then pedestrians). If a road's turn lane is GREEN and the pedestrian signal occurs along the same road, then the turn lane shall cycle to RED, and the pedestrian lights shall cycle to DON'T WALK. If there is waiting traffic in both roads of the intersection, then the intersection shall cycle as if it were in Mode 2 until no signals are received for the cycle times associated with the signals.

The same parameter set is used as in Mode 2, except the system responds to them with Mode 3 behavior, when Mode 3 is the selected mode.

Mode 4: Adaptive Mode

Mode 4 is for intersections with higher density traffic. In this mode, the intersection adapts to the local history of traffic by adjusting the cycle times depending on traffic density. This requires vehicle-counting behavior from the vehicle detectors. This is similar to Mode 2 and has the same parameters, plus:

Table A.5 Mode 4 parameters

Parameter	Value type	Description
Averaging Interval (AIp)	10 .. 120 minutes	(30) Period over which traffic volume is averaged to compute the relative density between the two roads
Minimum Density (MDp)	10 .. 1000	(100) Specifies the minimum number of vehicles, from both roads, that must have traversed the intersection before adaptive extension will be employed
90%	0 .. 60 seconds	(30) Length of time that a road's green time may be extended due to higher traffic volume when the road's traffic is 90% of total intersection traffic volume
80%	0 .. 60 seconds	(20) Length of time that a road's green time may be extended due to higher traffic volume when the road's traffic is 80% of total intersection traffic volume

Table A.5 Mode 4 parameters (continued)

Parameter	Value type	Description
70%	0 .. 60 seconds	(10) Length of time that a road's green time may be extended due to higher traffic volume when the road's traffic is 70% of total intersection traffic volume
60%	0 .. 60 seconds	(5) Length of time that a road's green time may be extended due to higher traffic volume when the road's traffic is 60% of total intersection traffic volume

The Vehicle Detector

Three types of vehicular detectors shall be supported: subsurface passive loop inductors (SPLIs), above-surface infrared sensors (ASIs) and above-surface radars (ASRs).

Subsurface detectors shall use a wired interface to communicate with the controller, while ASIs and ASRs shall support both wired and secure wireless communication. All vehicle detectors shall be able to perform vehicle counting.

In addition, ASIs and ASRs shall be able to receive directional transmissions from priority vehicle and emergency vehicle transmitters. The maximum range of such reception shall be no less than 250 feet and no more than 1000 feet.

Figure A.3 shows the relevant measures for both ASI and ASR detectors. When a vehicle enters the detection area (shown as the shaded area in the figure), the detector shall report the presence of a vehicle. Separate detectors are used for each lane in each direction.

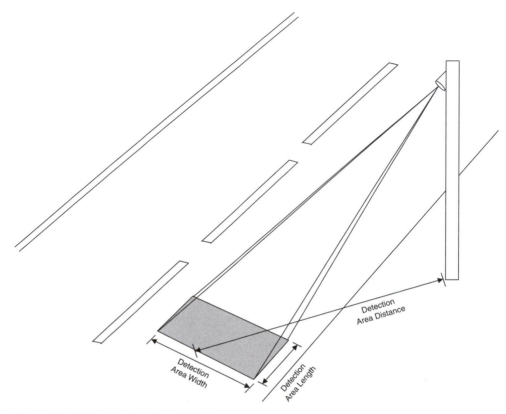

Figure A.3 Infrared and radar vehicle detector

Vehicular Traffic Light

Two kinds of traffic lights are supported: the standard three-light model and the 4-light model, the additional light being for a green turn arrow. When the intersection is allowing turn lane turns with the four-light model, then the green arrow shall be on.

If the traffic light can no longer detect that it is connected to the intersection controller, it shall FLASH RED. This shall occur in no longer than 10 seconds from a communications or RIC failure.

Pedestrian Light and Sensor

The pedestrian light is a two-signal light that can either be in the state of WALK, FLASHING DON'T WALK, or DON'T WALK, as shown in Figure A.5. If the light no longer detects that it is communicating with the RIC, it shall go to a state of DON'T WALK within 10 seconds of the communication or RIC failure.

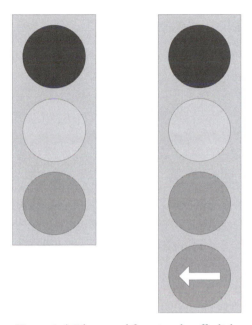

Figure A.4 Three- and four-signal traffic lights

Figure A.5 Pedestrian light

Front Panel Display

The front panel display (FPD) is an enclosed front panel display secured with a mechanical key access lock. The FPD provides an LCD for viewing information and keypad for entering information. The front panel display also has a wired Ethernet interface so that a local or remote network can view or set the values known to the system.

The FPD shall support the following choices as menus:

- Intersection Configuration Setup
- Mode 2 (Fixed Cycle Time) Setup
- Mode 3 (Response Cycle) Setup
- Mode 4 (Adaptive Cycle) Setup
- Intersection Statistics Display
- Device Manufacture Display

The FPD shall have a set of front panel keys and knobs as shown in Figure A.6. The turn knobs are used for menu selection and item selection. The numbers are for entering values; the arrow keys are for moving from field to field within segments of a parameter (such as mm:dd:yy for month:day:year date format). The Item Selection knob allows the selection of a parameter to change (if changeable). Before a parameter can be changed, it must be selected and then the user must press the EDIT key. Now the values can be entered with the keypad (if numeric) or selected from a list (if enumerated). When a change is made, it must be either confirmed by pressed the CONFIRM key or the change aborted with the ABORT key. The RESET key returns the value to the factory default if a current item is in the edit mode; or resets all parameters on the page to their defaults if no item is currently being edited.

The front panel has a power button that must be held down for 5 seconds before the power status can be changed. An LED shows RED if the controller is off but is receiving power, YELLOW if the controller is on but running off battery (UPS) or GREEN if on and operating from mains.

The FPD supports four Ethernet ports for external interfacing and digital I/O ports for interfacing with the lights and sensors. Software may be uploaded via the Ethernet power from a service technician's laptop.

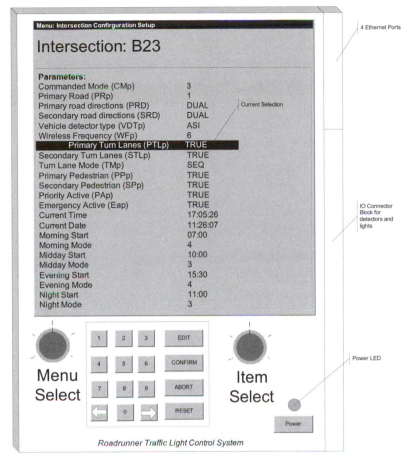

Figure A.6 Front panel display

The FPD supports secure wireless communication with up to 50 devices. Security includes 64-bit encryption of data and MAC address filtering. The wireless devices are primarily used to interact with infrared and radar vehicle sensors, but may also be used for connecting with a service technician's PC. The set up information for wireless, other than the frequency, is not available on the FPD and must be uploaded from the service technician's laptop. The range of the wireless connection shall be no less than 200 feet nor more than 1000 feet with an unobstructed line of sight.

Remote Communications

The RIC provides wired 10/100 Ethernet interfaces for remote monitoring and management (see the FPD). All parameters available on the front panel may be requested or set via this interface. In addition, traffic statistics may be viewed or reset via this interface.

Power

All components of the RIC that require power will accept 220–240 volts. Uninterruptible power supply (UPS) option is available to provide 1-hour or 10-hour functionality in the absence of line power. When the power is on, the UPS battery shall recharge. When running on UPS, the power light on the front panel display shall be RED. When running on main power, but charging the battery (and battery is at less than 90% capacity), the power light shall be AMBER. When on main power and battery is at 90% or more capacity, or when on main power and the UPS option is not present, the power light shall be GREEN.

The Coyote Unmanned Air Vehicle System (CUAVS) Specification

Overview

The Coyote Unmanned Air Vehicle System (CUAVS) is a system solution to medium-range reconnaissance in hostile environments with limited attack capability. It is a medium-range, long-endurance UAV system that can carry a variety of payloads to assist in ground, air, and sea operations. A full CUAVS consists of four Coyote Unmanned Air Vehicles (CUAVs) and a ground Coyote Mission Planning and Control system.

Primary CUAV System Components

The Unmanned Air Vehicle (UAV)

The Coyote UAV is meant to be a multipurpose reusable UAV with multimission capability. It is meant to operate at altitudes of up to 30,000 feet with ground speeds of up to 100 knots (cruise) and 150 knots (dash) and carry a payload of up to 450 lbs for durations in excess of 24 hours. The Coyote is meant to fly unimpeded in low-visibility environments while carrying either reconnaissance or attack payloads. While controllable from the ground station CMPCS, it is also capable of flying complex flight plans with specific operational goals of systematic area search, ground route (road-based) search, and orbit surveillance of point targets. Coupled with manned control from the ground, the Coyote provide sustained 24-hour flight with real-time visual, infrared or radar telemetry, with target recognition preprocessing. Communications are jam-resistant, although need not be anti-jamming in a high ECM environment. Control commands shall be encrypted while telemetry data can be compressed but unprotected. Telemetry rates for visual telemetry support 30 frames-per-second (fps) at 640 × 400 resolution. Range of flight is meant to be fully supported within line of sight (LOS) range but since the Coyote also has the ability to be passed among different CMPCSs, its range is considerably greater than LOS.

For navigation, the Coyote has on-board global positioning system (GPS) based navigation as well as being directly controllable from the ground station.

Unlike many smaller UAVs, the Coyote does not require specialized launch and recovery vehicles. It can use a short runway for either automated or remote-controlled takeoff and landing.

The Coyote Mission Planning and Control System (CMPCS)

Mobile CMPCS with capability to control up to four UAVs with a manned control station per UAV that fits into a smaller towable trailer. Each control station consists of two manned substations, one for controlling the CUAV and one for monitoring and controlling payloads. If desired, both functions can be slaved together into a single control substation. Control of the aircraft shall consist of transferring navigational commands which may be simple (set altitude, speed, direction), operational (fly to coordinate set, orbit point, execute search pattern, etc.), planned (upload multisegment flight plan) or remote controlled with a joystick interface. Stable flight mechanics shall be managed by the aircraft itself but this can be disabled for remotely controlled flight.

The CMPCS displays real-time reconnaissance data as well as maintaining continuous recording and replay capability for up to 96 hours of operation for four separate CUAVs. In addition, with attack payloads, the Coyote can carry up to four Hellfire missiles with fire-and-forget navigation systems.

Coyote Payloads

The Coyote UAV shall be easily configured to handle any of several different surveillance and reconnaissance payloads (video, forward-looking infrared [FLIR], radar), countermeasures (electronic countermeasures [ECM] and electronic counter-countermeasures [ECCM]) and attack payloads (Hellfire missiles). The payloads are primarily controlled from the ground, but certain functionality, such as search for target, can be assigned to the UAV itself. In this latter case, the Coyote performs any of a set of systematic search patterns for a target that matches the target specifications (such as tank, ship, convoy, platoon, soldier, missile launch platform, building) upon which time it notifies the controlling CMPCS.

The coyote can be configured with up to 450 pounds of payload and within that limitation can contain a mixture of surveillance, ECM/ECCM and attack payloads.

The Coyote Datalink Subsystem (CDS)

The CDS is used to communicate between the CUAV and the CMPCS. All commands to the CUAV shall be encrypted to prevent hostile intervention in the mission, while due to bandwidth limitations, real-time telemetry may be compressed but not encrypted. The CDS is an LOS communication system that is jam resistant, although not necessarily anti-jamming. It involves both spread-spectrum and high-speed frequency-hopping modes to improve jam resistance.

While the CDS is itself LOS, the CUAV may be passed among distributed CMPCSs. It can fly preprogrammed flight plans for up to 5 hours between CMPCS, so that it has extended range beyond LOS.

Detailed Requirements

The Unmanned Air Vehicle (UAV)

The UAV shall be no more than 30 feet long with a wingspan of no more than 10 feet. The UAV shall be able to take off and land using a runway of no more than 1500 feet by 50 feet, both automatically and via remote piloting. When fully loaded (450 lb payload), the system shall be able to maintain at least 24 hours of cruise flight. Gross weight shall not exceed 2100 pounds. The CUAV shall be able to fly reliably in inclement and low visibility weather; the CUAV shall be flyable without loss of reliability with 15-knot cross winds, 30 knot head winds, and in rain, sleet, hail, snow and in icy conditions.

The CUAV shall have a range of up to 400 nautical miles with an empty weight of no more than 1200 pounds (2100 pounds loaded). It has a ceiling of 30,000 feet and a 100-gallon (665 pound) fuel capacity.

The CUAV shall have a deployable parachute for emergency recovery.

Flight Modes

The CUAV is capable of fully automated flight, but it is envisioned that it will normally be under navigational control of the CMPCS. The following flight modes are supported:

- Fully automated
- Operational
- Remote (CUAV maintains stable attitude)
- High-fidelity remote (remote pilot maintains stable attitude)

In fully automated mode, the CUAV can take off, fly a preprogrammed mission and return. If the CUAV experiences an unexpected loss of communications with its CMPCS in excess of 60 minutes, it can abort the mission and perform an automated return and landing. This is meant to decrease incidence of UAV loss due to ECM and communications failures. If unable to land automatically, the CUAV will deploy its emergency parachute.

Automated navigation shall be performed using either intertial navigational system or a GPS system, or a combination of the two. The CUAV shall report its position along the CUAV control datalink at least every 3 seconds; this data shall include lat, long, and altitude information. When flying in environments where fewer than three GPS satellites will be visible, the CUAV shall have an on-board altimeter as well.

Mission Modes

Beyond flight modes, CUAV shall be designed for highly flexible mission parameters. Normal mission modes include:

- Preplanned reconnaissance
- Remote controlled reconnaissance
- Area search
- Route search
- Orbit point target
- Attack

A mission can consist of any number of sequential submissions, each operating in a different mission mode, depending on the current payload.

The Coyote Mission Planning and Control System (CMPCS)

The CMPCS is housed in a 30 × 8 × 8 triple-axis trailer that contains stations for pilot and payload operations, mission planning, data exploitation, communications, and SAR viewing. The CMPCS connects to multiple directional antennae for communication with the CUAVs, All mission data is recorded at the CMPCS since the CUAV has no on-board recording capability. The CMPCS has a UPS that can operate at full load for up to 4 hours in addition to using commercial power or power generators.

A single CMPCS can control up to four CUAVs in flight with one station per CUAV. Each CUAV control station provides both pilot and payload operations with separate control substations, although both functions can be slaved to a single substation for low-vigilance use.

For the reconnaissance payloads, the CMPCS shall provide enhanced automated target recognition (ATR) capability for all surveillance types—optical, infrared, and radar. While the CUAV has a rudimentary capability, the CMPCS provides much more complete support for the quick identification of high-value targets in the battlefield. This capability is specifically designed to identify mobile and time-limited targets that may only be exposed for brief periods of time before they go back into hiding. The system is expected to provide high clutter rejection and a low false-positive error rate. The ATR shall be able to identify and track up to 20 targets within the surveillance area, with likely identification and probability assessments for each. In addition to the ATR, the payload operator can add targets visually identified from reconnaissance data or gathered from other sources. The battlefield view can be transmitted over links to remote command staff for tactical and strategic assessment.

Communications shall include radio, cellular, and landline phones in addition to radio-based communication with the CUAVs.

The Coyote Reconnaissance Sensor Suite Payload (CSSP)

The CUAV is meant to be a mobile operational framework that is highly customizable for various missions. To that end, the CUAV can use gimbaled optical sensors (with zoom and spotter lenses), gimbaled FLIR sensors or synthetic aperture radar (SAR) for surveillance. In more situations, the sensor's positions are controlled by the CMPCS but may be automatically controlled for automated search missions. Telemetry shall be broadcast over the high-bandwidth surveillance datalink with the ability to transmit 640×400 resolution data at a rate of 30 fps. This data may be compressed but there is no requirement to encrypt the high-speed data.

The CUAV is equipped with a color nose camera (generally used by the air vehicle operator for flight control), a day variable aperture TV camera, a variable aperture infrared camera (for low light/night) and a synthetic aperture radar for looking through smoke, clouds or haze. The cameras produce full-motion video and the synthetic aperture radar produces still-frame radar images. The daylight variable aperture or the infrared electro-optical sensor may be operated simultaneously with the synthetic aperture radar. To reduce weight, the CUAV need not have all subcomponents of the CSSP but will normally have at least one.

The Coyote Hellfire Attack Payload (CHAP)

The CHAP provides the ability for the CUAV to carry and deploy up to four AGM-114 Hellfire missiles with the intention of being primarily for antitank and fixed target operations. The Hellfire missile is a solid propellant rocket with a maximum speed of 950 knots that weighs up to 100 pounds. The system shall be able to carry from 0 to 4 missiles within the limitation of payload weight. The CHAP, when present, shall include a laser sight designator with backscatter immunity that targets the missile prior to release or a radar target designator. Once released, the missile shall be self-guided, so the CHAP is intended for both guided and fire-and-forget operation. The missile shall be able to defeat reactive armor and successfully deploy in adverse and hazy weather. Warheads shall be high-explosive copper-lined antitank and antibunker.

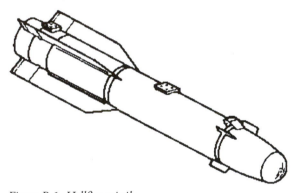

Figure B.1 Hellfire missile

The missiles attach to the CUAV on firing rails mounted on pylons, allowing two missiles to be mounted on each wing. Prior to release, target data is downloaded to the missile from the CUAV. Note that the CUAV cannot initiate Hellfire release itself; such control shall be relegated to the CMPCS.

When laser targeting, the missile is fired in a Lock-On-Before-Launch (LOBL) mode, in which the target is designated prior to missile release. The CUAV shall pulse the laser designator with a particular pulse pattern that is downloaded into the missile and is used to identify it from other laser designators that may be in use. Once located, the missile homes in on the reflected laser light. This operational mode requires the CUAV to maintain the laser designator on target until the missile reaches its target, so a single target can be fired upon at a time.

For fire-and-forget operations, the Hellfire missiles may be fitted with a radar seeker. Once the CUAV has located the target and the CMPCS has given the command to release, the missile maintains radar lock on the target. This allows the CUAV to fire and mask, or to seek other targets while the missile flies to its target. This allows up to four separate targets to be fired upon in rapid succession.

The Coyote Datalink Subsystem (CDS)

The CDS allows uploading of mission parameters, such as CUAV payload configuration, flight plans, communication frequency-hopping schedules, and so on; control of the vehicle operation pre-, in-, and post-flight; control of the payloads including sensor position, zoom level and selection; downloading of vehicle status, including current position, direction, speed, and fuel status; arming and releasing of missile payloads; setup and configuration of the operational flight program (OFP); and control of initiation of test modes.

The CDS consists logically of two distinct datalinks, although these may share the same communication media. The control datalink is an encrypted secure low-bandwidth datalink that supports vehicle and payload commands and status information. The required data rate is no more than 100 bits per second (bit/s) with a reliability of 0.9994 in an ECM free environment and 0.90 in a high ECM environment. The system shall be able to detect and correct all single- and dual-bit errors and detect multiple-bit errors with a reliability of 0.9994.

The high-speed data link is used for transmission from the CUAV only and is meant to include real-time telemetry, reconnaissance and surveillance data. The data link may compress the data, but it shall be able to transmit without information loss 640×480 optical data at a rate of 30 frames per second (fps) in an ECM-free environment. In a high-ECM environment, data rate shall be 320×200 at 15 fps for optical and FLIR imaging. SAR data shall be supported as single-frame images (640×400) transmitted at a rate not exceeding 1 fps. The high-speed data link need not be encrypted, as its reception by the enemy cannot lead to loss of the CUAV and operational experience has shown that real-time surveillance data is of little practical use to the enemy when intercepted.

UML Notational Summary

UML Diagram Taxonomy

The various kinds of diagrams defined in the UML 2.0

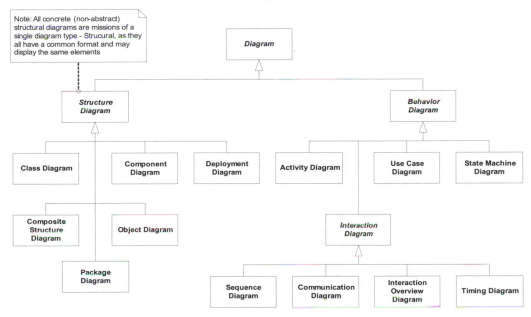

Common Diagram Elements

All diagrams may have a frame, name, and contain notes and constraints

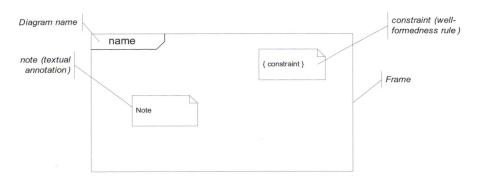

Class Diagram

Shows the existence of classes and
relationships in a logical view of a system

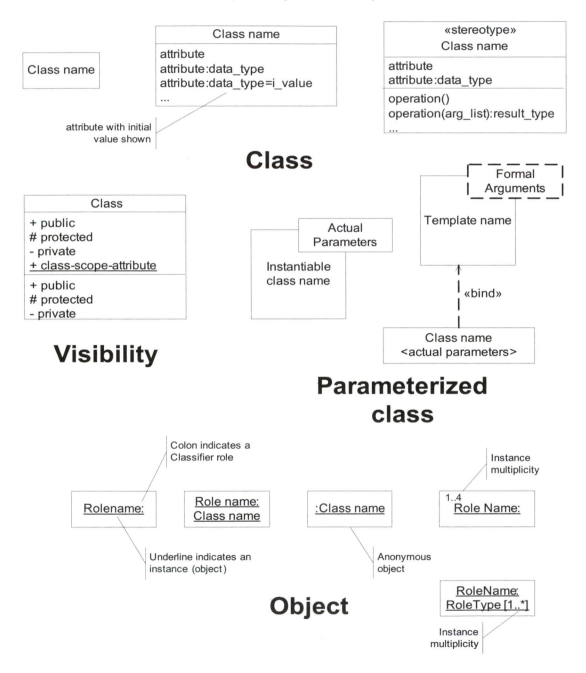

Class

Visibility

Parameterized class

Object

Class Diagram

Shows the existence of classes and
relationships in a logical view of a system

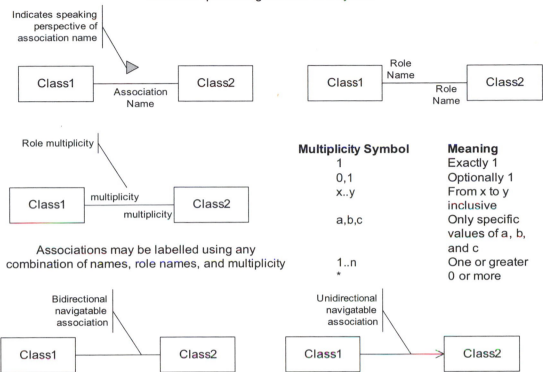

Indicates speaking perspective of association name

Association Name

Role Name

Role Name

Role multiplicity

multiplicity

multiplicity

Multiplicity Symbol	Meaning
1	Exactly 1
0,1	Optionally 1
x..y	From x to y inclusive
a,b,c	Only specific values of a, b, and c
1..n	One or greater
*	0 or more

Associations may be labelled using any
combination of names, role names, and multiplicity

Bidirectional navigatable association

Unidirectional navigatable association

Association

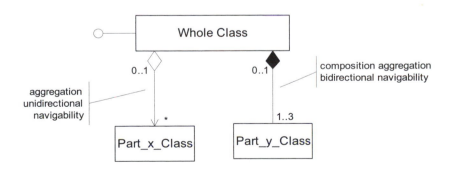

aggregation unidirectional navigability

composition aggregation bidirectional navigability

Aggregation and Composition

Class Diagram

Shows the existence of classes and
relationships in a logical view of a system

Advanced Associations

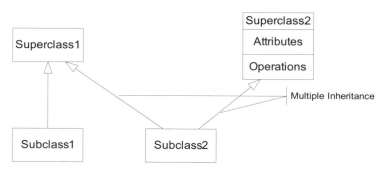

Generalization

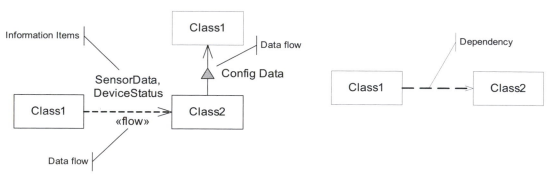

Dependency and Data Flow

Class Diagram

Shows the existence of classes and
relationships in a logical view of a system

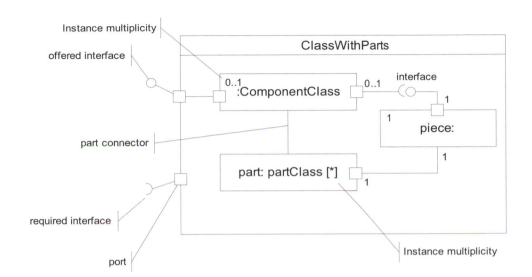

Structured Classes

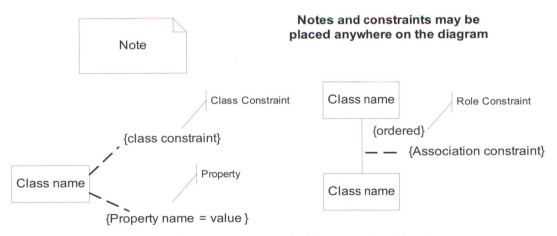

**Notes and constraints may be
placed anywhere on the diagram**

Notes and Constraints

Class Diagram

Shows the existence of classes and
relationships in a logical view of a system.

«stereotype»

<<stereotype>>

Standard stereotype
indicators

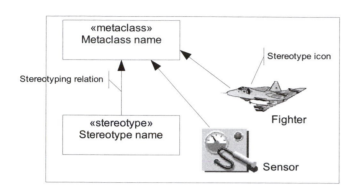

«metaclass»
Metaclass name

Stereotyping relation

«stereotype»
Stereotype name

Stereotype icon

Fighter

Sensor

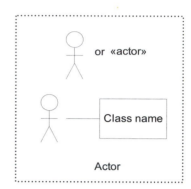

or «actor»

Class name

Actor

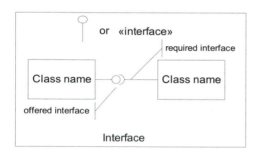

or   «interface»

required interface

Class name

Class name

offered interface

Interface

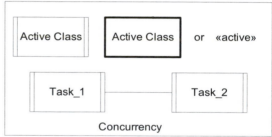

Active Class   Active Class   or   «active»

Task_1   Task_2

Concurrency

Stereotypes

Communication Diagram

Shows a sequenced set of messages illustrating a specific example of object interaction.

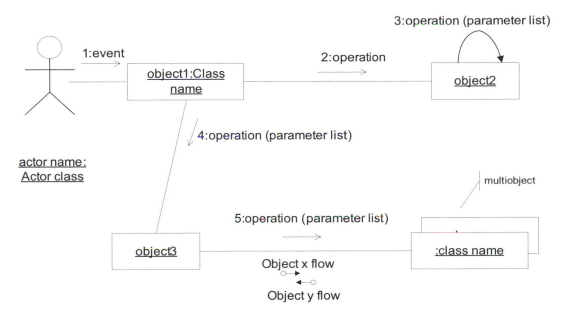

Object Collaboration

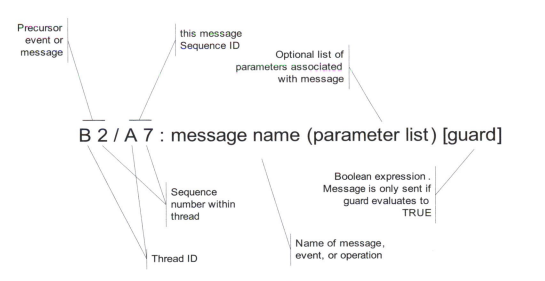

Message Syntax

Sequence Diagram

Shows a sequenced set of messages illustrating a specific
example of object interaction.

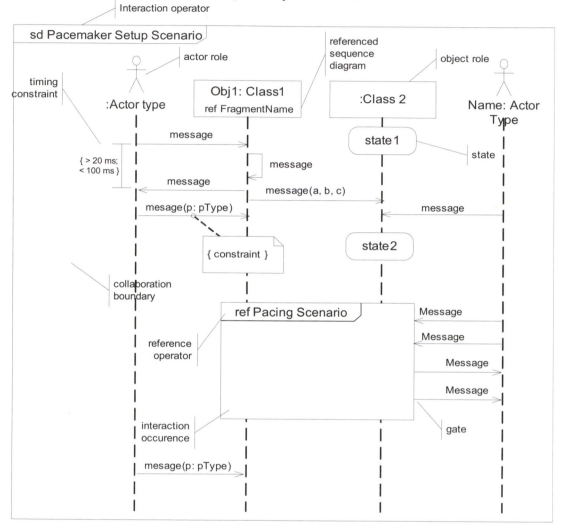

Sequence Diagram Elements

Sequence Diagram

Shows a sequenced set of messages illustrating a specific
example of object interaction.

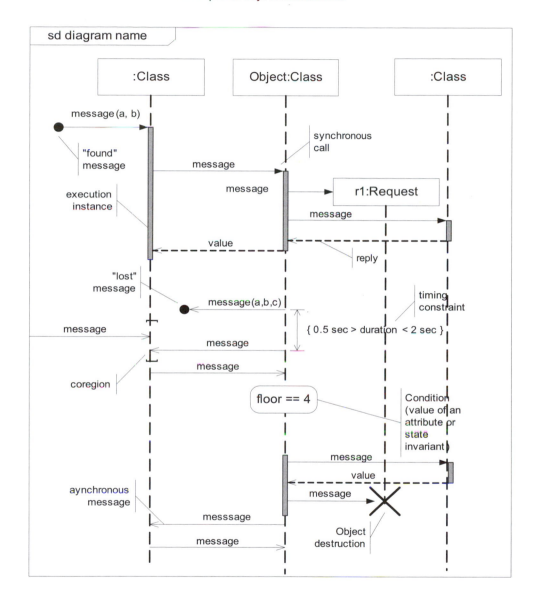

Sequence Diagram Annotations

Sequence Diagram

Shows a sequenced set of messages illustrating a specific
example of object interaction.

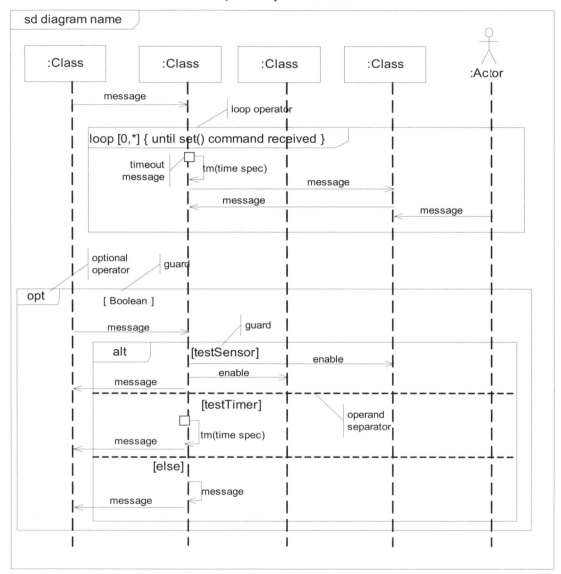

Interaction Operators

Sequence Diagram

Shows a sequenced set of messages illustrating a specific
example of object interaction.

Operator	Description
sd	Names an interaction fragment.
alt	The alt operator provides alternatives, only one of which will be taken. The branches are evaluated on the basis of guards, similar to statecharts. An "else" guard is provided that evaluates to TRUE if and only if all other branch alternatives evaluate to FALSE.
opt	Defines an optional interaction segment; i.e., that may or may not occur.
break	Break is a short hand for an alt operator where one operand is given and the other is the rest of the enclosing interaction fragment. A sequence diagram analogue to C++ "break" statement.
loop	Specifies that an interaction fragment shall be repeated some number of times.
seq	Weak sequencing (default). Specifies the normal weak sequencing rules are in force in the fragment.
strict	Specifies that the messages in the interaction fragment are fully ordered - that is, only a single execution trace is consistent with the fragment.
neg	Specifies a negative, or "not" condition. Useful for capturing negative requirements.
par	Defines parallel or concurrent regions in an interaction fragment. This is similar to alt in that subfragments are identified, but differs in that ALL such subfragments execute rather than just a single one.
criticalRegion	Identifies that the interaction fragment must be treated as atomic and cannot be interleaved with other event occurrences. It is useful in combination with the par operator.
ignore/consider	The ignore operator specifies that some message types are not shown within the interaction fragment, but can be ignored for the purpose of the diagram. The consider operator specifies which messages should be considered in the fragment.
assert	Specifies that the interaction fragment represents an assertion .

Interaction Operators

Use Cases

Use cases show primary areas of collaboration between the system and actors in its environment. Use cases are isomorphic with function points.

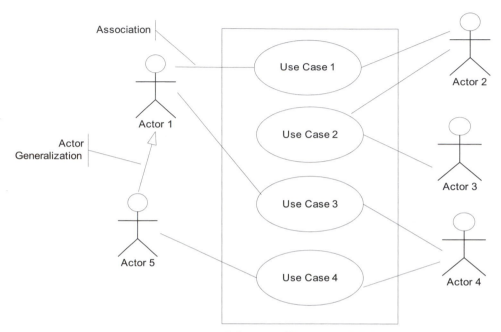

Use Case Diagram

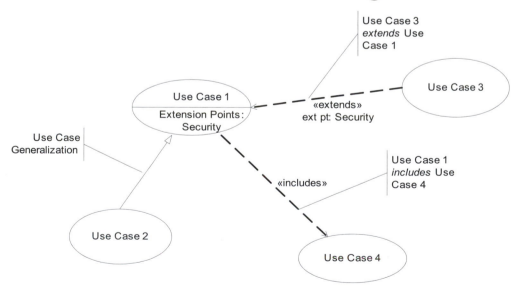

Use Case Relationships

Implementation Diagrams

Implementation diagrams show the run-time dependencies and packaging structure of the deployed system .

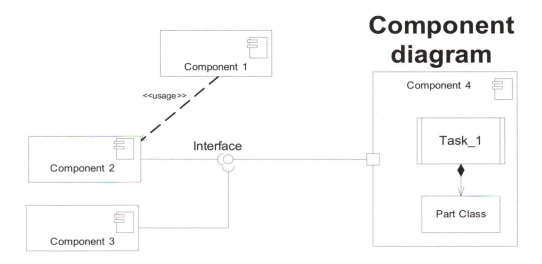

Component diagram

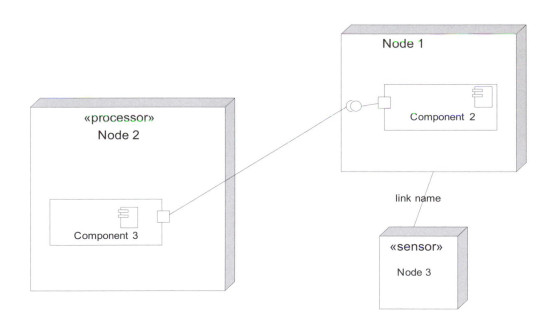

Deployment Diagram

Packages

Shows a grouping of model elements. Packages may appear within class and other diagrams.

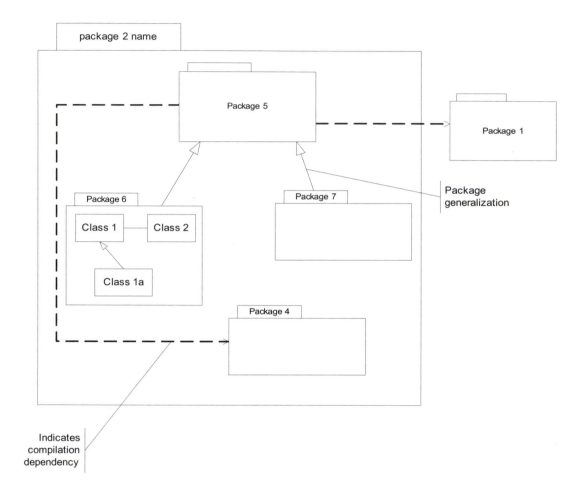

Statechart

Shows the sequences of states for a reactive class or interaction during its life in response to stimuli, together with its responses and actions.

State name

entry / action-list
event-name:action-list
do / activity-list
defer/ event-list
...
exit /action-list

State icon

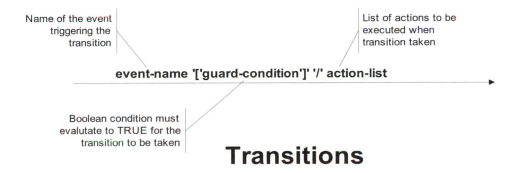

Name of the event triggering the transition

List of actions to be executed when transition taken

event-name '['guard-condition']' '/' action-list

Boolean condition must evalutate to TRUE for the transition to be taken

Transitions

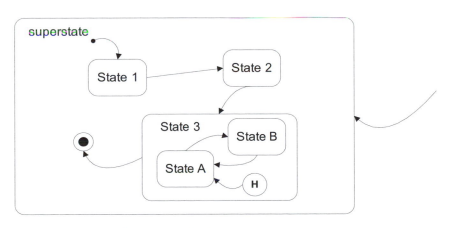

superstate

State 1

State 2

State 3

State B

State A

H

Nested States

Statechart

Shows the sequences of states for a reactive class or interaction during its life in response to stimuli, together with its responses and actions.

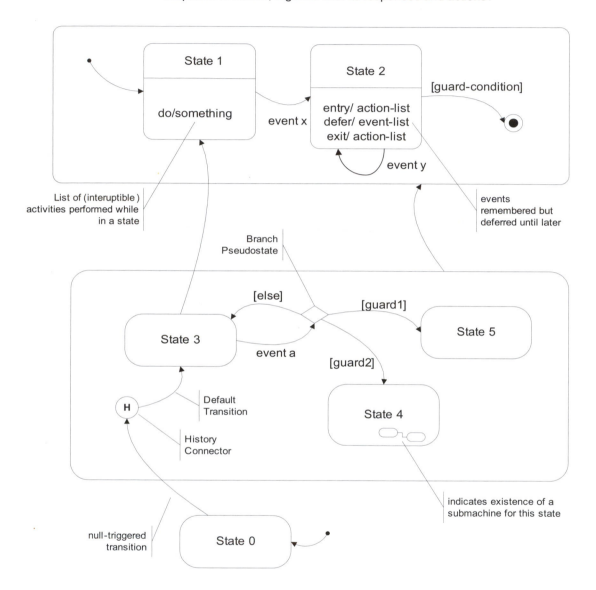

List of (interuptible) activities performed while in a state

events remembered but deferred until later

Branch Pseudostate

[else]

[guard1]

State 5

event a

[guard2]

Default Transition

History Connector

State 4

indicates existence of a submachine for this state

null-triggered transition

State 0

Sequential substates

Statechart

Shows the sequences of states for a reactive class or interaction during its life in response to stimuli, together with its responses and actions.

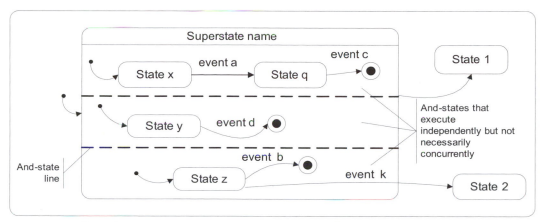

Orthogonal substates (and-states)

Symbol	Symbol Name	Symbol	Symbol Name
Ⓒ or ◇	Branch Pseudostate (*type of junction pseudostate*)	Ⓗ	(Shallow) History Pseudostate
Ⓣ or ◉	Terminal or Final Pseudostate	Ⓗ*	(Deep) History Pseudostate
		•↘	Initial or Default Pseudostate
⊢ Fork	Fork Pseudostate	●	Junction Pseudostate
⊢ Join	Join Pseudostate	●	Merge Junction Pseudostate (*type of junction pseudostate*)
○ [g] [g]	Choice Point Pseudostate	○ label	Entry Point Pseudostate
		⊗ label	Exit Point Pseudostate

Pseudostates

Statechart

Shows the sequences of states for a reactive class or interaction during its life in response to stimuli, together with its responses and actions.

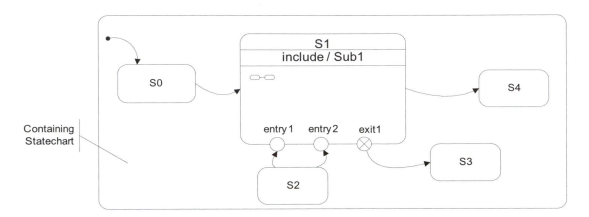

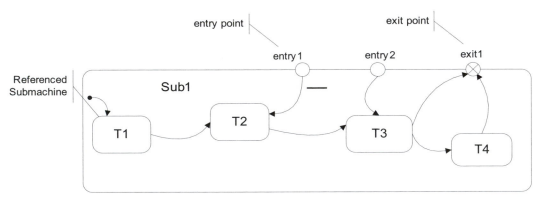

Submachines

Activity Diagrams

Activity Diagrams are a behavioral diagram based on token flo semantics and includes branching, forks, and joins.

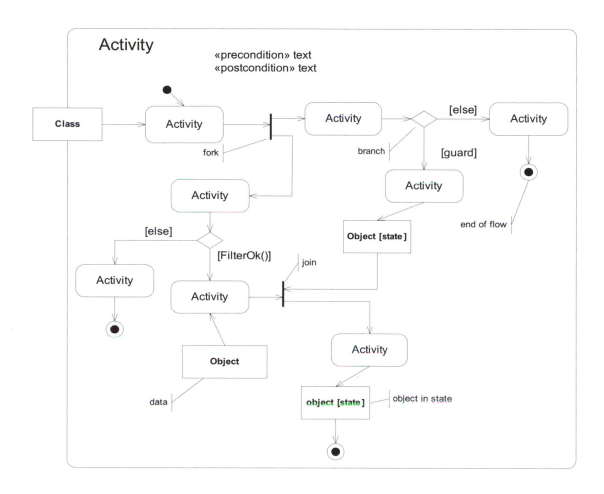

Activity Diagrams
Activity Diagrams are a behavioral diagram based on token flo semantics and includes branching, forks, and joins.

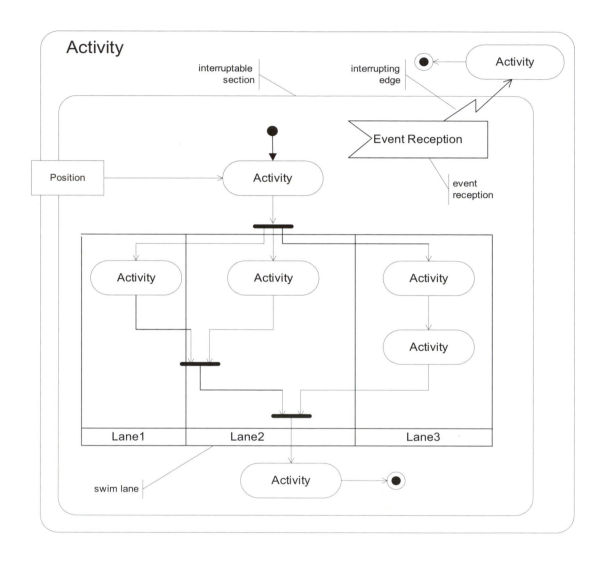

Timing Diagrams

Timing diagrams show the explicit change of state or value along a linear time axis
(Timing Diagrams are new in the UML 2.0 standard)

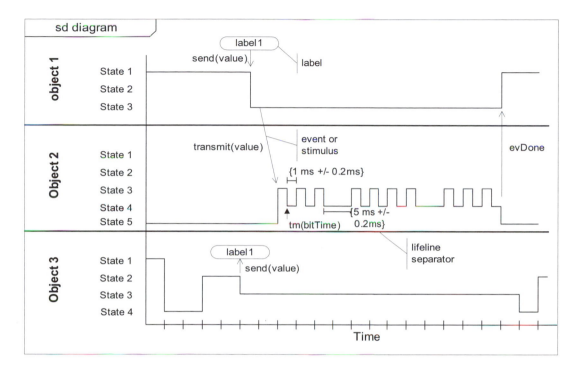

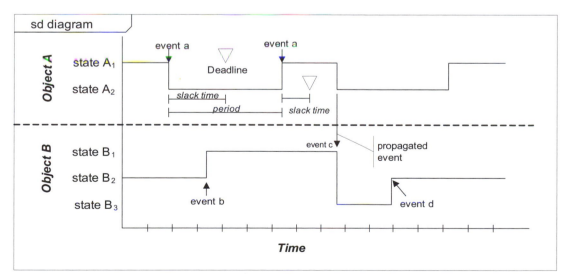

Timing Diagrams

Timing diagrams show the explicit change of state or value along a linear time axis
(Timing Diagrams are new in the UML 2.0 standard)

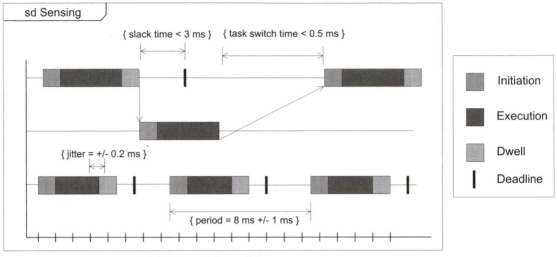

With Shading

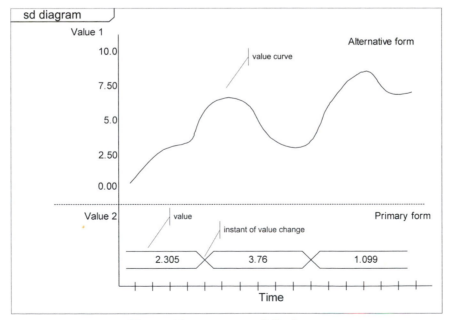

Continuous Values

Index